3D Printing

3D Printing

3D Printing
Technology, Applications, and Selection

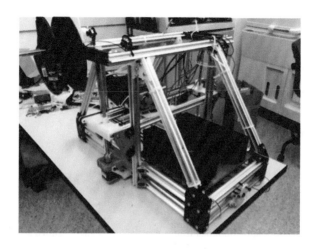

Rafiq Noorani
Loyola Marymount University

CRC Press
Taylor & Francis Group
Boca Raton London New York

CRC Press is an imprint of the
Taylor & Francis Group, an **informa** business

CRC Press
Taylor & Francis Group
6000 Broken Sound Parkway NW, Suite 300
Boca Raton, FL 33487-2742

First issued in paperback 2020

© 2018 by Taylor & Francis Group, LLC
CRC Press is an imprint of Taylor & Francis Group, an Informa business

No claim to original U.S. Government works

ISBN-13: 978-1-4987-8375-0 (hbk)
ISBN-13: 978-0-367-78196-5 (pbk)

Visit the Taylor & Francis Web site at
http://www.taylorandfrancis.com

and the CRC Press Web site at
http://www.crcpress.com

This book is in memory of my late parents (Osman Ali and Nurunnessa Noorani) and my teachers, who taught me the importance of hard work, honesty and lifelong learning.

The book is dedicated to my family: Zarina (wife), Sabrina (Daughter), Alejandro (son-in-law), and Layla and Elana (my two beautiful grandchildren).

Contents

Book Organization

The book is organized into 10 chapters. The objectives of each chapter are described below:

- *Chapter 1: Introduction:* This chapter introduces the readers to the world of rapid prototyping and 3D printing (3DP). Definition, historical development, important developments, and applications are described.

- *Chapter 2: How Does 3D Printing Work?* This chapter provides an introduction to 3DP and rapid prototyping (RP) as part of additive manufacturing. It provides a difference between conventional manufacturing and 3DP. It also describes the basic steps in making prototypes-solid model, data conversion, building the part, software limitations, and a detailed case study.

- *Chapter 3: Design of a 3D Printer:* This chapter provides the details of the construction of a 3D printer. The goal of this chapter is to use available kit components to make an inexpensive 3D printer in-house that can be as fast as the commercial fused deposition modeling printer.

- *Chapter 4: Calibrating the 3D Printer:* This chapter describes the calibration of the printer that is vital for fast, easy, and quality prints. It is the mind of the printer that controls the printer. An open-source software, called *Marlin*, has been used to control all aspects of the printer.

- *Chapter 5: Materials for 3D Printing:* RP materials play a key role in the quality, strength, and accuracy of prototypes. This chapter reviews 3DP materials (polymers, metals, ceramics, and composites) from the fundamental approach by reviewing chemical bonding. It discusses the current 3DP materials and processes and gives advantages and disadvantages of each material.

- *Chapter 6: Classifications of RP and 3DP Systems:* Most printers fall into the categories of liquid-, solid-, and powder-based printers. This chapters describes some of the most common but successful 3D printers on the market today.

- *Chapter 7: Scanning and Reverse Engineering:* In this chapter, the principles and applications of scanners and reverse engineering (RE) are introduced and discussed. RE plays an important role in shortening product lead time. This chapter discusses the background, scanning techniques, data conversion systems, and future trends in RE.

- *Chapter 8: Common Applications of 3D Printers:* The applications of 3DP are mind-boggling. This chapter introduces readers to the common household can opener to designing and printing a flower vase, recreating human face, and medical applications in making human fingers.
- *Chapter 9: 3D Printing in Medicine:* This chapter describes medical imaging and specific applications aspects of 3DP, and also presents 3D printing of a C1 Vertebrate from CT scan data.
- *Chapter 10: How to Select RP and 3DP?* This chapter describes the guidelines for implementing 3DP and RP in the workplace, operations and management issues, and service bureaus. It describes the development of an expert system that can be used to select a particular type of 3DP and RP that suites the needs of the user.

There is enough material in this book to cover a single-semester course. Each instructor can select those portions of the book that cover the topics he or she wants to cover in a particular course. Students can also use the book as reference for other topics that are not covered during the course.

Preface

3D printing (3DP), also known as additive manufacturing (AM) and rapid prototyping (RP), is a technology that takes information from a computer-aided design and "prints" it on a 3D printer, which creates a solid object by building up successive layers of material. The technology was developed in 1989 and, since then, has been expanding exponentially. Figure 1 shows the explosive growth of 3D printers in recent years. 3D printers are similar in many ways to RP machines except that they are cheaper (less than $5000.00), smaller, and sometimes, less accurate in terms of strength and surface finish. Commercial systems, such as stereolithography and fused deposition modeling, have passed into public domain. These are the big companies who have been making big and expensive RP machines for a long time. However, inventors and entrepreneurs around the world are using commercially available kit components to build 3D printers whose prices have fallen from $100,000 to over $500. The author of this book and two of his students have recently designed and built a 3D printer at Loyola Marymount University. The cover page shows the completed printer. The 3DP essentially consists of a frame, extruder, electronics, materials, and software. This project has been a great experience for the author and the students. Now that the printer has been built, the next challenge is the continuous improvement of the printer for its precise control.

The advantage of the digital design is that it can be tweaked easily with a few mouse clicks. The 3D printer can run unattended and make many things, which are too complex for a traditional manufacturing process. In time, these amazing machines may be able to make almost anything, anywhere, anytime. The applications of 3D printer are especially amazing. From 3D printed organs and prosthetics to tools that have been printed in

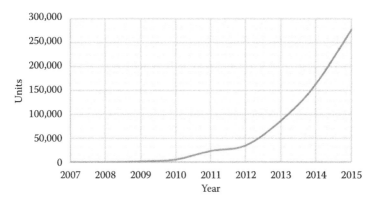

FIGURE 1
Growth of 3D printers.

zero gravity on the International Space Station, not to mention some of the incredible sculptures that redefine what is possible to achieve in a physical medium, 3D printers are being used almost in every industry to transform the way we do things and lead us into a bold new future. 3D printer was even touted by President Obama as having potential to "revolutionize the way we make almost everything" during his 2013 State of the Union address.

Those of us who are engaged in design have been using RP since its inception in 1989. It is a powerful tool that has many advantages. Some of the benefits attributed to RP and 3DP are (1) reduced lead-time to produce prototypes, (2) improved ability to visualize the part due to its physical existence, (3) early detection and correction of design errors, and (4) increased capability to compute mass properties of components and assemblies.

An impressive number of individuals, educational institutions, manufacturing organizations, and industries have discovered RP, 3DP, and their potential benefits. Many high schools, colleges, and universities have purchased and implemented RP machines and 3D printers into their curricula. Students are routinely taught the principles and applications of 3DP. Students use 3DP tools in their laboratory and design/capstone projects. Automobile manufacturers and suppliers are using both 3DP and 3DP to prototype various mechanical components. Aerospace companies mostly use RP machines to produce parts and assemblies for both commercial and military aircraft, as well as in the product development cycle. Manufacturers of business machines use RP and 3DP to prototype computer parts, copier, telephone, and fax machine parts.

The primary purpose of this book is to present the technology, applications, and selection of 3D printer as it relates to the world of concept development, design, and manufacturing. The book is designed for hobbyists, entrepreneurs, high school students, and college/university students for an entry-level course on 3DP in design and manufacturing automation. The book can also be used as a supplement to computer graphics text and manufacturing texts that are currently being used. The book is also intended to serve as a reference for industrial and academic practitioners interested in competitive product design and manufacturing.

MATLAB® is a registered trademark of The MathWorks, Inc. For product information, please contact:

The MathWorks, Inc.
3 Apple Hill Drive
Natick, MA 01760-2098 USA
Tel: 508-647-7000
Fax: 508-647-7001
E-mail: info@mathworks.com
Web: www.mathworks.com

Acknowledgments

I have been blessed by having many talented colleagues, friends, and students during the 35 years in which I have been actively involved in computer-aided design and computer-aided manufacturing (CAD/CAM) and additive manufacturing (AM). Many ideas in this book have originated from close friendships, coauthorships, collaborative projects, and formal and informal discussions at various conferences.

Support from the National Science Foundation has been invaluable in developing the CAD/CAM laboratory and Rapid Prototyping laboratory at LMU. It is the second NSF grant that I received in 1995 that helped me buy the first rapid prototyping (RP) machine and start my work on RP. I am grateful to Parsons Foundation who gave me a grant of $150K in 1997 to purchase my second RP machine and a coordinate measurement machine.

As an international professor, I have worked very closely with Dr. Toshihiro Ioi of Chiba Institute of Technology and Dr. Hideyuki Kanematsu of National Institute of Technology, Suzuka College in Japan, and Dr. Kwang-Sun Kim of Korea University of Technology and Education (KoreaTech) in Cheonan. These professors have invited me to give lectures on RP and 3D printing for their classes, and industries. I am very grateful to these professors for their continued support and encouragement.

I am very grateful to my first reviewers who made corrections and improvements to the chapter contents. I am thankful to my students, Michael Lynch, Patrick Hodgkiss, Cassandra Jacobsen, Cole Merritt, Chris Mitts, Matt Fumo, Max Yarema, Andrew Dominguez, Alex Hendricks, Abdullah Alrashdan, Faisal Aldhabib, Kelly Tivoli, and others who worked with me very closely to develop the chapters of this book. I must acknowledge the encouragement I have received from my wife, Zarina Noorani, to complete the book.

I am grateful to Loyola Marymount University for allowing me to use university resources in the preparation of the manuscript.

Finally, I would like to thank CRC Press for turning this book into a reality. I am very thankful to Cindy Carelli, my acquisition editor, Renee Nakash, editorial assistant, and Mathi Ganesan of Codemantra, who worked with me very closely to convert the manuscript into a textbook.

About the Author

Rafiq Noorani received his BSc degree in mechanical engineering from Bangladesh University of Engineering and Technology (BUET), a master of engineering degree in nuclear engineering, and a PhD in mechanical engineering from Texas A&M University.

He is currently a tenured professor and graduate program director of mechanical engineering at Loyola Marymount University in Los Angeles, California. He previously taught engineering at Texas A&M University, the University of Louisiana at Lafayette, and Gonzaga University in Spokane, Washington before joining Loyola Marymount University in 1989. His teaching interests include CAD/CAM, rapid prototyping, 3D printing, and robotics. He has received over 35 external grants totaling over $2 million including six grants from National Science Foundation (NSF). He has published 12 journal papers and 80 conference papers in national and international journals and conferences. With the help of a recent NSF grant, he has worked with his colleagues to develop an undergraduate program in nanotechnology at LMU.

Dr. Noorani published the first textbook on rapid prototyping in North America in 2006 titled: *Rapid Prototyping—Principles and Applications*. The book was published by John Wiley & Sons. He also became a *Fellow* of American Society of Mechanical Engineers (ASME). According to ASME guidelines, the fellow grade of membership recognizes exceptional engineering achievements and contributions to the engineering profession.

As an international professor, he teaches and lectures on CAD/CAM, 3D printing, and robotics in many countries including Japan, South Korea, and Bangladesh. Dr. Noorani is a professional member of American Society of Mechanical Engineers (ASME), Society of Manufacturing Engineers (SME), and American Society for Engineering Education (ASEE).

1

Introduction

1.1 Introduction

Mastering the art of 3D printed components, sub-assemblies, and products is vital for any product development process. Over the last few years, new and exciting technologies have emerged that are changing the ways that products have been launched. 3D printing (3DP) is an example of such technology, which is revolutionizing the design and manufacturing of new products. This technology is being used to fabricate physical solid models for early verification of concepts, e.g., form, fit, and function as well as reducing lead-times for product development [1].

Not long ago, 3DP was only associated with prototyping. Nowadays, 3DP technology has come a long way from basic rapid prototyping (RP) machines of the 1980s, and 3DP is more than just prototyping with today's materials and technology. 3DP process offers transformative advantages at every part of the manufacturing process from initial concept design to production of final products to be marketed, and all the steps in between. There is extremely large variety of 3D printers and technologies available these days; therefore, it is important to be thoroughly informed in order to choose the right 3D printer for specific applications.

First, 3DP technology as we know it today emerged in 1980s, and a thought of having a personal mini 3DP workstation was a dream. Even just a few years ago, in-house 3DP technology was enjoyed only by a very few professional design engineers, and the applications were very limited to creation of concept models and simple prototypes. Once considered a luxury, 3DP has proven to yield long-term strategic value by enhancing design-to-manufacturing capabilities and reducing the time to market final products. 3DP technologies have allowed an ever-growing number of creators, designers, engineers, physicians, researchers, academics, and manufacturers to unleash all the benefits of RP/3DP in-house across the entire design process.

Leading manufacturing companies harvested the full power that 3DP technology has to offer and are now using 3DP to evaluate more concepts in short time to improve the decision-making and design process early in product development. As the design process moves forward, technical decisions are iteratively tested at every step to guide decisions, to achieve maximum performance, minimize manufacturing costs, and deliver the highest quality. In the preproduction phase of manufacturing process, 3DP allows accelerated first concept product development and assists with troubleshooting of any problems that may arise. In the final production stage of manufacturing process, 3DP allows for higher productivity, increased flexibility, reduced logistics costs, economical optimization, improved product quality, reduced product weight, less overall parts in the assembly, and greater efficiency in a growing number of industries [2].

In this opening chapter, 3DP will be defined. In addition, study of 3DP history and the importance of this technology for product development will be covered as well. Applications of this technology through a case study and future trends will also be discussed.

1.2 The World of 3DP

The field of 3DP encompasses a wide variety of new methods, technologies, and applications that have already stimulated some fascinating research. Many companies have found exciting new ways to improve product development processes and enhance profitability. The prospect of being a part of this new technology with its promise of radical improvement in the way business is done should be as highly motivating to the reader, as it is to the author. In this section, some definitions and applications of 3DP will be discussed, and examples of some companies that are using 3DP will be given.

1.2.1 What Is 3DP?

3DP, as described in this text, refers to the fabrication of a physical, 3D part of arbitrary shape directly from a numerical description (typically a CAD model) by a quick, totally automated, and highly flexible process without any tooling. According to Wohler's Report 2014, 3DP is defined (as also defined by ASTM International Committee F42) as "fabrication of objects through the deposition of a material using a print head, nozzle, or other printer technology." However, the term is often used synonymously with *additive manufacturing (AM)*. AM is a broader term that encompasses building 3D physical models. Prototypes, patterns, and tooling components using various materials such as plastics, ceramics, metals, etc. [3].

1.2.2 What Is RP?

RP is a technology that also refers to the fabrication of 3D part of an arbitrary shape directly from a numerical description by a quick, totally automated, and flexible process. RP is the term that has been used since 1989 when the RP process was invented. From 1989 to 2005, the additive technology was known as RP. After nearly two decades of the use of RP, the personal 3DP revolution started with an open-source project known as RepRap. Since then, the growth of 3DP has been phenomenal, exceeding over 72,000 units sale in 2013. The difference between RP and 3DP is minimal and the two terms can be used interchangeably. In particular, 3DP is associated with machines that are lower in relative price (<$5000) and overall functional capability. Both RP and 3DP are part of AM.

As it was mentioned before, 3DP is a relatively new technology that creates profound effect on the product development process of design and manufacturing industries worldwide. 3DP technology can prototype parts very rapidly in most cases, i.e., in hours rather than in days or weeks. This technology is quickly expanding to include rapid tooling (RT) and rapid manufacturing (RM). RP, tooling, and manufacturing are technologies that are also widely used these days. 3DP can be sometimes referred to as *desktop manufacturing, direct CAD manufacturing,* and *instant manufacturing.*

The unique characteristic of 3DP manufacturing process is that it makes physical prototypes one layer at a time. Therefore, following terms that emphasize this layer-by-layer manufacturing characteristic are sometimes used as well: *layered manufacturing, material deposit manufacturing,* and *material addition manufacturing.*

The last group of terms for 3DP emphasizes the words "solid," "freeform," and "fabrication." This group of terms includes *solid freeform fabrication* and *solid freeform manufacturing.* The word "solid" refers to the final solid state of the material, although the initial state may be solid, liquid, or powder. The word "freeform" stresses the fact that 3DP can prototype complex shapes with little or no constraint on shape form. 3DP is also related to "automated fabrication," which describes new technologies for generating 3D objects from computer files in a completely automated process.

1.2.3 The History of 3DP

Charles Hull, the founder of 3D Systems, is consider as the founder of 3DP. He obtained his first patent for 3DP in 1984 for his stereolithography apparatus

(SLA), which uses UV light to cure photopolymer resin in a vat to make prototypes. The principle of 3DP is that a software can slice a 3D object into layers of particular thickness and a machine can stack them together one above the other, forming the 3D part. In the early 1980s, the computer-aided design (CAD) was still in its infancy, and Mr. Hull had a challenge of how to translate the CAD file in a file format that the 3D printer can interpret and print. Again, with the help of a consulting firm (Albert Consulting Group), he developed the stereolithography (STL) file format that could be used by any 3DP machine. Although people talk about some limitations of the STL file, today, it is still the *de facto* standard for 3DP process. We shall talk more about this STL file in the next chapter [4].

Scott Crump, the founder of fused deposition modeling (FDM) printing process, developed the FDM technology in 1989 and founded the Stratasys company. Crump also developed acrylonitrile butadiene styrene (ABS) materials for the FDM machine that is widely used by a vast majority of 3D printers today. 3D Systems and Stratasys are the two biggest 3DP companies in the world. Over the years, these two companies have acquired other companies with different technologies and have truly become the best and biggest 3DP companies in the world.

Dr. Carl Deckard and Dr. Joe Beaman developed selected laser sintering (SLS) process at the University of Texas in the mid-1980s independently of the SLA and FDM processes. While SLS and FDM were successful in making plastic and nylon parts, none of the technology could make prototypes in metals. SLS was the only technology at that time to make prototyping parts in metals. The technology was originally sold by DTM Corporation, which was later purchased by 3D Systems.

In the early 1990s, the Massachusetts Institute of Technology (MIT) invented inkjet 3D printing, known as 3DP. The license for 3DP was given to Z Corp. In the early 2012, 3D Systems bought the company to acquire all the associated patents and licenses.

Many of the industrial patents of 3D Systems and Stratasys expired a few years ago. Especially, the expiration of FDM technology started the consumer 3DP technology, as we know it today. The 3DP revolution was started in 2005 with the open-source project called RepRap. The purpose of the RepRap project is to make a machine that can replicate itself. With the participation of entrepreneurs and hobbyists, the 3DP revolution is going on in full swing. These 3D printers are very inexpensive (<$5000) but very robust in most cases. Most 3D printers use FDM and stereolithography technology. They use open-source hardware and software to operate. With the advancement of computer technology and electronics and software, these 3D printers are the future. The historical growth of 3D printers since 2005 bears testimony to this effect. Some of the most popular consumer 3D printers of the market are MakerBot Replicator and FormLabs 1+. We shall describe some of the most popular 3D printers in the next few sections of this chapter.

1.2.4 Applications of 3DP

There are many types and classes of physical prototypes, but their main purpose is to minimize risk during the product development process. Some of the specific applications of 3DP technology are mentioned below:

- Communication of product characteristics
- Engineering concept definition
- Form, fit, and function testing
- Engineering change clarification
- Client presentations and consumer evaluations
- Bid proposals and regulation certification
- Styling, ergonomic studies
- Facilitate meeting schedule and making milestones
- Masters for silicone rubber tooling
- Masters for spray metal tooling (all processes)
- Masters for epoxy tooling to be used for injection molding
- Master/Pattern for investment casting
- Tooling for injection molding

1.2.5 The Basic Process of 3DP

There are many different 3DP processes available today that use a variety of materials, such as wax, plastics, metals, and many more. The most widely used and popular 3DP processes are SLA, selective laser sintering (SLS), and FDM. Although there are many different 3DP processes, almost all of the processes operate by forming solid objects, using either liquid, powder-based, or solid materials. This means that all these 3DP techniques follow the same basic process for part creation. This basic 3DP process consists of the following steps [5]:

1. Create a CAD model of the design.
2. Convert the CAD model to STL file format.
3. Slice the STL file into 2D cross-sectional layers.
4. Grow the prototype.
5. Clean and finish the model.

The first step in the process is to create the CAD solid model using CAD software packages, such as AutoCAD, Pro/Engineer, or Solid Works.

In the second step, CAD file of the part to be manufactured is converted into STL file format. Because various CAD software packages use different algorithms to represent solid objects, STL file format has been selected as the de facto standard in the RP industry. The STL file represents a 3D surface of an object as planar triangles. The file contains the coordinates of the vertices and the direction of the outward normal of each triangle. The STL file format is the best file format to represent all surfaces, in preparation for the "slicing" algorithms. More about the STL file format will be discussed in Chapter 2.

The third step involves "slicing" of the STL file using a proprietary software program, provided by the manufacturer of the 3DP machine in which the model is to be produced. The preprocessing software imports the STL file, and lets the user orient the part, and adjust the size and slice thickness of the model. The layer thickness may vary from 0.01 to 0.7 mm depending on the capabilities of the machine. Lower slice thickness increases the accuracy of the prototype but also increases the time to build it. The preprocessing software may also generate a structure to support portions of the model during its buildup. Supports are necessary for creating features such as overhangs, internal cavities, and thin-walled sections. 3DP processing software also provides information about how much time and material will be required to make the prototype.

The fourth step involves the actual making of the prototype. Once the STL file is processed and saved, it is sent to the 3DP machine. During this part of 3DP process, the machine essentially acts as a printer by building the prototype one layer at a time. Most modern 3DP machines can operate unattended once the initial setup is completed, and the printing operation is initiated.

The final step is removing the part from the machine and cleaning it before use. This step is also known as postprocessing. It also involves postcuring of photosensitive materials, sintering powder materials, and removing of the support materials. Some prototypes are also subjected to surface treatment, such as sanding, sealing, or painting to improve their surface finish, appearance, and durability.

1.2.6 Industries Using 3DP

Design and manufacturing industries use 3DP technology to reduce the manufacturing time and for marketing their products, and more importantly to cut manufacturing costs. The early uses of prototypes were mainly for visualization to check form, fit, function, and the early verification of design error. 3D prototypes are now used for many other purposes. Functional models and fit/assembly together represent approximately 48.5% of all 3D models used. About 20.4% of 3D printed models are used for patterns for prototype tooling and metal casting. This area of application is expected to increase dramatically in the near future. Figure 1.1 shows how companies are using 3D printed models. Figure 1.2 shows the same data as Figure 1.1 but in terms of the number of units produced for each application of 3D printed models [3].

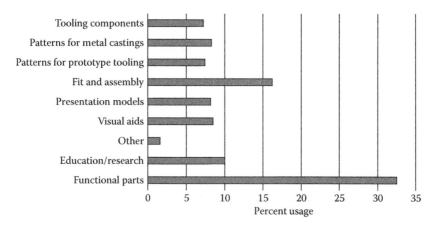

FIGURE 1.1
Purposes of 3D printed models.

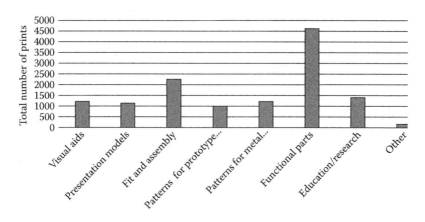

FIGURE 1.2
Number of 3D printed models used per specific application area.

1.3 Growth of RP and 3DP Systems

1.3.1 Growth of RP

While the RP industry continues to grow, the rate of growth has been very flat for the years 1997–2002 due to the lethargy in the overall economy. In spite of this, during 2000–2001, established commercial 3D printer manufacturing companies released new technologies, new materials, and new applications for 3DP machines. Overall, the electronic component and costs have decreased, while computing power increased. This made sales of commercial

3D printing systems (RP Systems) to skyrocket. Figure 1.3 shows the growth of commercial RP systems since 1988.

While the development of commercial 3DP systems was taking place, and the sales were increasing, personal 3DP machines that could be easily used in the office environment are becoming more mainstream. Improvement in computer numerical control (CNC) software and hardware allowed to scale down 3D printers to a personal level. Research and development for new materials for personal 3D printers allowed personal 3D printers to become a reality. Figure 1.4 shows the growth of personal 3D printer sales from 2008 to 2014. As it can be seen from the plot, the growth is exponential, and over the span of 6 years, sales grew from 355 units in 2008 to 72,503 units in 2014. If this trend continues, soon

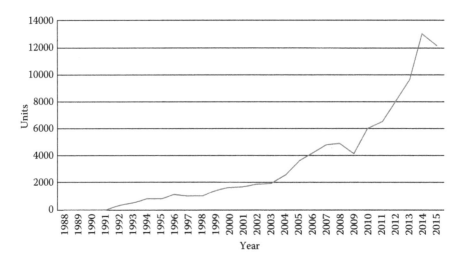

FIGURE 1.3
The growth of RP technology.

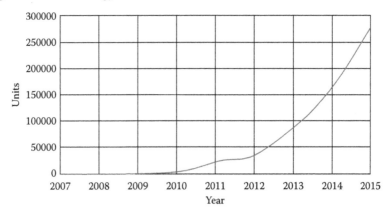

FIGURE 1.4
The growth of personal 3DP technology.

personal 3D printers will be in every household, office, etc., and these machines could be used to manufacture things quickly and inexpensively.

1.3.2 3DP Sales by Regions

The United States of America no longer leads in the production and sales of professional-grade, industrial 3DP systems, as shown in Figure 1.5. Israel now holds the top spot with 41.1%. When Stratasys merged with Israeli-based Objet in December 2012, the newly formed entity chose to register as a company of Israel. This development created a very big difference. In 2012, the U.S. produced nearly 61% of all industrial systems. In 2015, it declined to

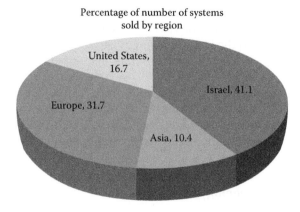

FIGURE 1.5
Industrial 3DP systems sold by region.

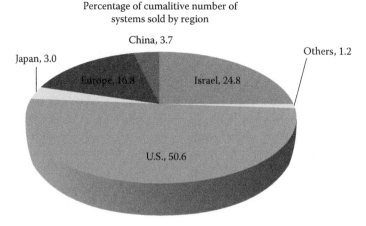

FIGURE 1.6
Cumulative total number of 3DP systems sold by geographic region.

an estimated 16.7%. Europe's position grew from 19.2% in 2012 to 31.7% in 2015. Asia's share grew less rapidly, from 5.4% to 10.4%.

The chart in Figure 1.6 provides the cumulative total number of industrial AM systems sold from each geographic region beginning 1988 through 2015. The U.S. System manufacturers are responsible for 50.6% of all industrial AM machines sold over this period, which is down from 71.2% three years earlier. Israel's share has increased from 10.0% to 24.8% while Europe has increased from 11.5% to 16.8%.

1.3.3 Units Sold by U.S. Companies

Among the U.S. manufacturers of RP equipment, 3D Systems has been the leader in the industry since the beginning. Since its acquisition of DTM System, it became an even more dominant company in the United States of America With combined sales of 415 units sold in 2001, 3D Systems controls 40% of the unit sales in the U.S., followed by Stratasys, which sold 277 units. Figure 1.7 shows the percentage of the market share based upon units sold by the U.S. companies in the year 2015. For cost-conscious companies, all major RP companies have developed low-cost RP machines, such as ThermoJet (3D Systems), Z400 (Z-Corp), and Dimension (Stratasys). The low-cost machines are like concept modelers that are used for concept verification, form, fit, function, and early verification of design errors and are compatible with an office environment.

1.3.4 3D Printer Technology Development

3DP is a technology that is used by most industrialized countries of the world today. The general goals, objectives, and needs for 3DP are the same in the United States, Europe, and Asia; however, the emphasis may be different for different countries. In Japan, the emphasis on accuracy is the predominant

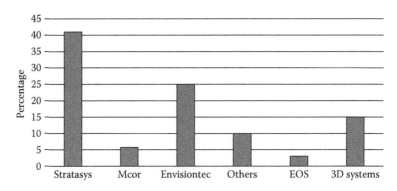

FIGURE 1.7
Market share percentage based on units sold by U.S. companies.

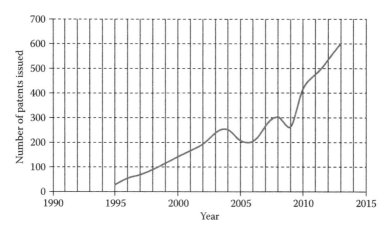

FIGURE 1.8
Increase in the number of patents in 3DP technology.

consideration for using 3D printer technology, while in Europe, there is more emphasis on developing prototypes in metal components and tooling. As mentioned before, the 3DP technology was developed in the United States, and the United States is still the leader of this technology, both in terms of its development and applications.

3DP is not a technology by itself, rather, it draws on other technologies such as computer-aided design (CAD), computer-aided manufacturing (CAM) and CNC. However, nowadays, 3DP is highly integrated with all the supporting technologies mentioned earlier. 3DP could now be considered as the technology by itself. Even though there has been a lot of new developments in this area, this technology is in a continuous improvement state. In this section, the technological development of 3DP technology will be briefly discussed.

The historical developments of 3DP technology can be understood easily through tracing the list of major patent developments. Many of the latest patents have been classified due to the use of 3DP technology in aerospace and defense industries. Only patents that are not classified could be shown.

The growth of the issued patents in 3DP technology increases steadily year by year. The plot in Figure 1.8 shows the growth in issued patents since 1995. In 2013, the number of issued patents related to 3DP technology continued to increase linearly.

1.4 Current Popular 3D Printers

In this section, a series of new technologies that are available in the market are reviewed. These technologies allow "automatic" production of prototype parts directly from a solid model. The systems discussed here are commercial systems currently available and most popular in the United States of America [6].

1.4.1 MakerBot Replicator 2

The Replicator 2 is very attractive based on its appearance and its features such as an onboard control with a large LCD screen that allows untethered printing directly from an SD card without the necessity of connecting the printer to a computer. Some of its upgraded features from last year's design include its acceleration, better quality on high-detail prints, and reduction of noise. It has also been renovated to slice faster and accurately for improved print quality. In addition, an "auto-layout" option is now available for automatic arrangement of models. The total cost of such machine, as shown in Figure 1.9, is around $2199 with a print volume of 11.2″ × 6″ × 6.1″. It does not contain a heated bed, open-source software, or hardware. The print material is known to be polylactic acid (PLA) with OS supported by Linux, Windows, and Mac. The printer control software is MakerWare with slicing software of MakerBot slicer.

1.4.2 FormLabs 1+

The system uses a UV laser beam that is steered by a mirror galvanometer system to selectively cure the resin to form each layer. As shown in Figure 1.10, the machine contains a matte aluminum chassis, OLED display,

FIGURE 1.9
MakerBot Replicator 2.

FIGURE 1.10
FormLabs 1+.

bright and orange cover, and a single button interface. The orange cover filters out the blue light that cures the resin because without it the results would be a semisolid chunk of resin. The platform at the upper surface of the orange casing is where the part will be built upside-down by pulling the part from the resin. The container at the bottom is where the resin is located. The process of printing with the Form 1+ is as simple as it seems while also containing two connection cables: one power and one USB. The file is simply to be loaded and hit Print or Form. An attractive feature of the system is that it contains PreForm software, which detects whether the model contains errors and helps orienting it correctly for printing.

The system was tested using the bundled clear resin with a low resolution of 100 microns. Although, it offers three different resolutions of 100, 50, and 25 μ, the lowest was chosen in order to have as many prints as possible. Many samples were printed of different size and duration in which the results concluded that the Form 1+ was excellent because the parts printed were beautiful. Its only flaw is that the prints stuck really well to the platform and requires force to remove it. Therefore, a small toolkit has been added with the printer for the removal process. The price of the machine is $3,299 with a 1 l resin and a build volume of 4.9″ × 4.9″ × 6.5″. The machine is OS supported by Mac and Windows. The printer is not completely untethered, but a computer can be disconnected after printing begins. It does not contain an open-source hardware or software. Lastly, its printer control software and slicing software is PreForm.

FIGURE 1.11
Ultimaker 2.

1.4.3 Ultimaker 2

A top-quality 3D printer, such as the one in Figure 1.11, was built so that it would be easy to use among professionals and beginners. Similar to the Replicator 2, it is attractive based on its appearance and features on the onboard control interface. The system includes a sleek aluminum-polymer panel and frosted acrylic. The material used to print can be ABS or PLA, in which the prints are easily removed. The Ultimaker 2 uses the FDM, which makes it simple to use and has the ability to print in a variety of materials.

The price of this machine is over $2000 including VAT at iMakr. The printer is OS supported by Linux, Mac, and Windows. It is also an open-source hardware and contains Cura as a printer control software. The slicing software is called CuraEngine.

A comparison of the specifications of the three printers listed above can be seen in Table 1.1.

TABLE 1.1

3D Printer Summary

Parameters	FormLabs 1+	Ultimaker 2	MakerBot Replicator 2
Weight (kg)	8	11	11.5
Build volume	125 mm × 125 mm × 165 mm	230 mm × 225 mm × 205 mm	285 mm × 153 mm × 155 mm
Cost of materials	$101.17 to $151.75 per liter	$35.41 per VAT	$29.99 to $49.99 for 1 kg
Cost of printer	$3820.71	$2042.56	$2499.00
Resolution	25–200 µ	20 µ	100 µ
Materials	Thermal resin	ABS or PLA filament	ABS or PLA filament

1.5 Applications in Education and Industry

The very first commercial 3DP machine made its debut in 1988, and very soon after that it became operational. Recently, 3DP industry has grown into an important and integral part of the new product development process. The use of 3DP technology has decreased time to market, decreased initial development costs, and improved product quality by giving design and manufacturing, and engineering the tool to verify and fine tune initial designs before committing them to expensive tooling and fabrication processes for final manufacturing. In order to achieve all the goals to their most optimal point, it is also vitally important to select a proper type of printer for the job.

1.5.1 Choosing the Right 3D Printer for Specific Application

With the variety of 3DP systems available today, and ever-growing 3DP market, choosing the right 3D printer among all the possible alternatives can be a daunting task. There are significant differences between different 3DP machines and technologies. Today's 3D printers can use a variety of materials as edible food to metals and carbon fiber, and anything in between. Also, mechanical properties, definition of the features, surface finish (roughness) environmental resistance, visual appearance, accuracy and precision, useful life, thermal properties, and more parameters of 3D printed parts are greatly influenced by the 3DP technology that was used to create the part. Before considering investment in3D printer, it is important to first define the primary applications where 3D printer will be used in order to guide the selection of the right 3DP technology that will provide the greatest positive impact on the final manufactured product. Therefore, the four most common applications of 3DP technology will be described, and a general guide for the selection of appropriate 3D printer for the job will be developed.

Concept models improve the early design decisions that impact every subsequent design and engineering process. Selecting the correct design path will reduce the cost later in the product development, and will overall shorten the entire development cycle. This allows to market products quicker. 3DP is the ideal way of initial design evaluations, any possible alternative designs, and enables cross-functional input. During the early part of the design process, it is desirable to evaluate real physical concept designs and any possible alternative designs with models that look and feel like the real thing. However, these models do not to be fully functional at this stage. Hence, for most concept modeling application, the key performance characteristics to look for in a 3D printer include print speed, part cost, ease of use, and life-like print output.

Prototype Design Evaluation involves designs using solid modeling techniques to facilitate the design evaluation by providing such function as viewing, shading, rotating, and scaling. However, there is no substitute for a design evaluation where the design can be held and felt physically. Without the actual prototype, no design can be accurately visualized. The visualization of prototypes can help reduce design error and analysis of the product.

Functional prototypes are physical models/prototypes of the final product design that is used to test and verify all the functions of the product. As product designs mature, engineers and designers need to verify and test all design elements to ensure that the new product will function as designed. 3DP technology allows design verification to be an iterative process where designers identify and address design challenges to come up with improvements, or quickly identify pitfalls of the design, and solve these pitfalls. Some typical applications of functional prototypes are testing for form and fit, functional performance, assembly verification, kinematic performance, aerodynamic testing, etc. Verification prototypes provide real hands on feedback to quickly prove design theories through practical application. For verification purposes, the parts should provide a true representation of design performance. Material characteristics, model accuracy, feature detail resolution, and build volume are some of the most important characteristics of 3D printers to look for in choosing 3DP technology for functional verification purposes.

1.5.1.1 Preproduction Applications

As the product development cycle is finalized, attention is turned toward the manufacturing process. At this stage, it is common to make large investments in tooling, jigs, and fixtures that are needed for the manufacturing process. At this stage in the process, complex supply chains are created for material delivery, hardware purchases, tool purchases, etc. All these aspects of the manufacturing process will require lead times, and often these lead-times could be longer than anticipated. This, in turn, will lengthen the time to market the final product and will bring significant financial losses. Using 3DP for this application can, in a variety of ways, reduce the investment risk, shorten lead times drastically, and shorten the overall time for a final product launch. At this stage, the functional performance of 3DP materials is critical. Accuracy, precision, and repeatability are of great importance to ensure that the final product quality is achieved, and the manufacturing tooling will not require expensive and time-consuming rework.

1.5.1.2 Digital Manufacturing

3DP is often associated with freeform fabrication; therefore, some 3DP technologies can create virtually unlimited geometry without the restrictions of traditional manufacturing methods such as NC and CNC machining. This gives engineers and designers' greater freedom in design process in order to create the levels of product functionality. Manufacturing costs are reduced by eliminating time and labor-intensive traditional manufacturing processes, and reducing raw material waste. 3D printed parts in this application may be end-use parts or sacrificial product enablers, such as casting patterns, which simplify and streamline the manufacturing process. More and more industries are using 3D printers for end-use parts as 3DP technology quality becomes ever better, and material capability widens. For digital manufacturing applications, the key 3D printer characteristics are high accuracy, precision and repeatability, material properties, specialized paint materials specifically engineered for application requirements, part cost, and production capacity.

1.5.2 Applications of 3DP in Product Development

The world has already entered into a new era of global competition for providing products and services. Rapid acceleration of new and emerging technologies is fueling this growth in all aspects of business. Companies engaged in product development and manufacturing are in tremendous competition to bring a product to market faster, cheaper, with both higher quality and functionality. Reducing the timeline for product development saves money in the overall time-to-market scenario. 3DP is the technology that helps companies reduce the cycle of product development, and also facilitate making design improvements earlier in the process where changes are less expensive. The impact of 3DP on product development is shown in Figure 1.12.

Figure 1.12 shows a general design process for product development. This design process is carried out in the following order:

1. Preliminary (first iteration) design concepts
2. Parametric design
3. Analysis and optimization
4. Creation of prototype
5. Prototype testing and evaluation
6. Comparing to design criteria
7. Final product or prototype

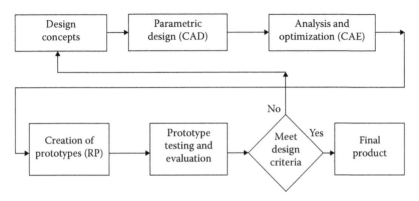

FIGURE 1.12
The impact of 3DP on product development.

The first step in the design process is the creation of preliminary design concept. This concept is a first iteration design that arises directly from customer's product requirements. It requires the engineer to use experience, knowledge, and ingenuity to work out a preliminary concept of what functions the design should be able to perform.

The second step involves the parametric design of the design concept. Because the design phase could be carried out with the help of computers, CAD techniques using software packages, such as AutoCAD, Pro/Engineer, SolidWorks, or any other commercially available CAD programs are used in this step.

Once a preliminary design is completed, this design is analyzed and optimized based on the results of the analysis. The optimized design is then ready to be prototyped/3D printed, tested, and a final preproduction part could be fabricated.

Design and manufacturing companies use 3DP technology to produce models, prototype injected molded components, and castings that are used in many applications, such as, home electronics, personal computers, cars, airplane parts, instrument panels, and medical applications. In addition, some of the other RP uses include the following: improving time to market, creating visualization tools, and reducing waste in the design process by identifying flaws in the design early in the manufacturing cycle.

1.5.3 Application in Reverse Engineering

Reverse engineering (RE) is the technique of reproducing the surface geometry of an existing physical part or model directly into a 3D data file format by using contact or noncontact 3D scanner. In many cases, only the physical model of an object is available, but there exists a need to create multiple copies of this object. Examples of such objects are handmade prototypes, craft works, reproduction of old engineering objects and sculptured bodies found in medical and dental applications. In order to facilitate 3DP of these physical models, it is essential to create their CAD model first. RE is the quickest way to get 3D data into digital CAD format [7].

1.5.4 Product Development Using a Low-Cost RE Process

1.5.4.1 Project Objective

The objective of this experiment was to reverse engineer a basketball using a low-cost RE method. This was carried out because RE is a new tool that expedites product development for which this design data exists. The objective was accomplished with 360 degree scanning, divided into eight sections with two additional top and bottom scans. The scans were merged, converted to a scan file, converted again to an STL file, and rapid prototyped with the FDM-1650 machine. The original and reverse-engineered basketballs were compared on texture and dimensional accuracy.

1.5.4.2 Procedure

The part desired to be replicated was marked with a white pen in specific locations around the surface of the basketball that were used as reference points. The part was divided into eight scans, completing the 360 degrees. The first scan is shown in Figure 1.13. The first two scans were taken, then, using the reference points, the scans were merged as shown in Figure 1.14.

This process was then repeated for the rest of the eight scans. Then, two additional 180 degree scans were performed on the top and the bottom of the basketball and merged with the model. Once all the scans were performed,

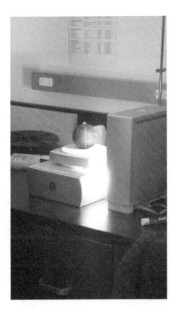

FIGURE 1.13
Scanner setup and first scan.

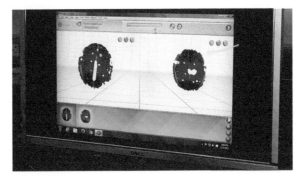

FIGURE 1.14
Merging the scans together.

the file was saved as a scan file and converted to an STL file. The file was opened and the part file was created using the FDM-1650 RP machine.

1.5.4.3 Results

The final scan detailed the entire basketball, including all eight side scans and the top and bottom scans. The rapid prototyped basketball shown in Figure 1.15 mimicked the original ball and produced the following dimensional analysis results, which are displayed in Table 1.2. A percent difference was also calculated to compare the two values.

1.5.4.4 Conclusion and Discussion

The low-cost method to produce a part without computer data was successfully completed. The deviation between the dimensional analysis of the two parts proved that the low-cost method is reliable and accurate. This is apparent with the low percent difference calculations with the largest deviation being 4.084% and the smallest being 0.133%. These results display that RE is a powerful tool that can be utilized to generate complex parts in which no

FIGURE 1.15
Rapid prototyped basketball compared to original basketball.

TABLE 1.2

Comparison of Actual and RE Diameters

Trial	Actual Diameter (in.)	RE Diameter (in.)	Percent Difference (%)
1	3.0090	3.0130	0.133
2	2.9500	3.0730	4.084
3	3.0265	3.0135	0.430
4	3.0435	3.0800	1.192
5	3.0985	3.1335	1.123

computer data exist. This experiment proved that there is an acceptable low-cost method that can be used to accurately produce a prototype that meets the designer's needs and specifications. Using the low-cost RE method, the basketball was successfully replicated with a low deviation in dimensional analysis.

1.5.5 Application in Casting and Pattern Making

Another important application of 3DP is in casting and pattern making. In this case, the pattern and the cores for the casting process are made using special 3DP processes. Investment casting, for example, is a very convenient process employed to make jewelry, dental fixtures, small mechanical components, blades for turbine engines, etc. Investment casting offers an alternative to machining, especially in cases where the shapes and details are extremely irregular. Investment casting allows fabricators to manufacture complex parts with ease. In investment casting, a model made of wax is dipped in a ceramic slurry. This creates a ceramic shell around the wax model (pattern). The wax is then melted out of the shell, the shell is hardened by heating and the molten metal of choice is then poured in. Metals used for this process include 14K and 18K gold, bronze, and silver, to list a few. For example, a bronze sculpture model was made using investment casting technique. Figure 1.16 shows this model and the original nonbronze model.

Biomet, a maker of surgical implants, has used FDM 3DP to produce 60 implantable casting designs per month. The company has also produced

FIGURE 1.16
Investment cast model in bronze and nonbronze model.

another 600 to 1000 castings monthly, using injection-wax patterns from steel tools, due to the volume requirements needed from a single design.

Engineers have tested many designs using ABS patterns, and they have been able to achieve accurate castings in as little as 2 weeks.

1.5.6 Application in RT

RT is a manufacturing process that combines 3DP technology with conventional tooling practices to produce molds quickly. RT typically uses 3D printed model as a pattern or uses the 3DP process directly to fabricate a tool for a limited volume of prototypes. There are three main types of RT: direct tooling, single-reverse tooling, double-reverse tooling, and triple-reverse tooling. With the increase of reversals, the durability of a product can be improved, but the product cost increases while its precision decreases.

The need for faster and less expensive tooling solution has resulted in more than 20 RT methods. Many more advanced RT methods, processes, and systems are being developed.

One such method is material forming, which was applied to sheet metal forming by Boris Fritz and Rafiqul Noorani. The objective of their research was to study the application of materials forming using RT as a part of RP. By using a hydraulic press and a rubber pad forming (Guerin process), sheet metal was formed into shapes. These shapes came from STL files that were made into molds that the metal was pressed into. The results of the project showed that RT can reduce the time of production for sheet metal forming from 96 h down to 2–4 h. This 96%–98% decrease in production time not only saves manufacturer's time but also cost. A sample of a sheet metal part created from a rapid prototyped mold is shown in Figure 1.17.

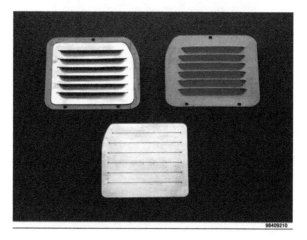

FIGURE 1.17
Sheet metal part (left) created from RP mold (right).

1.5.7 Application in RM

The current methods of RP and RT have profound effects on the manufacturing process. Some companies are extending the envelope of 3DP and RT into a process called RM. Using forward thinking, companies have successfully used 3DP and RT for the production of finished manufactured parts. Many believe that this practice of making final parts, which is RM, is the future of 3DP.

There are still a lot of challenges to RM such as the limitations of speed, materials, and accuracy of 3DP machines. However, as companies demonstrate more successful applications, more and more efforts will be put into the RM technology, and many of the obstacles to RM will be overcome in the next couple of years.

The aerospace industry is one of the biggest users of 3DP technology. However, the industry imposes stringent quality controls on all the prototypes used for various applications. Yet, Boeing's Rocketdyne division has successfully used 3DP and RT to manufacture hundreds of parts for applications on the International Space Station and Space Shuttle fleet [8]. 3D printed parts have also been used for F-18 fighter jets. Rocketdyne has manufactured these parts *using 3D Systems' SLS process.*

The students of the SAE Formula 1 team at Loyola Marymount University (LMU) routinely use rapid prototyped parts for their vehicle, and some are produced as final production parts for their cars built to compete in races. Some of the parts include intake manifold, side view mirrors, and various molds and models for the vehicle.

Another application of RM is in the *Invisalign* technology. *Invisalign* is the invisible new way to straighten teeth without braces. It uses a series of clear, removable aligners to gradually straighten teeth, without metal or wires. *Invisalign* uses 3D computer imaging technology to depict the complete treatment plan from the initial position to the final desired position from which a series of custom-made "aligners" are produced. Each "aligner" moves teeth incrementally and is worn for about two weeks, then replaced by the next in the series until the final position is achieved. These "aligners" are made by RP machines (mostly by SLA machines of 3D Systems).

1.5.8 Jet Engine Applications

Jet engines are very complex machines that consist of many extremely complicated parts. Using conventional manufacturing techniques such as CNC machining can be extremely expensive and time-consuming process to create jet engine components. In some cases, even CNC machining cannot cope with the complexity of the parts. 3DP is starting to revolutionize the manufacturing methods used for jet engine parts. It was a matter of time, and the technology being developed before 3DP started to be used in industries where precise engineering was of the biggest importance proving that the 3DP technology was evolving past the novelty status. The very first part that was 3D printed for jet engine application was T25 housing for a compressor inlet temperature sensor that was designed and 3D printed by general electric (GE) aviation. This part will be used as a retrofit for approximately 400 GE90-94B jet engines for Boeing 777-200 aircraft. While this part was the very first 3D printed part approved by Federal Aviation Administration (FAA), GE has plans to increase the number of 3D printed parts. One of the very interesting parts that are being studied for 3DP by GE is fuel nozzle. Figure 1.18 shows 3D printed T-25 sensor that GE aviation has developed. Traditional sensors are assembled from approximately 10 parts that have to be precision manufactured. Using 3DP allows manufacturing the entire assembly as a single piece without worrying about fit tolerances of parts. Also, 3DP techniques allow for more precise part features to be manufactured. So, why is GE pursuing 3DP for its jet engines? There are many benefits for both GE's bottom line and the environment, including lower cost, lighter weight, and better fuel efficiency. There is also the fact that the process produces far less waste than the traditional method of milling or cutting away material from a metal slab to produce a part. The processes lets engineers design, and build a single part that can replace a complicated assembly of several parts. Engineers are able to build a part layer by layer with metal powder and then fuse it all together with an electron beam of laser. This allows them to create more complex and precise parts than would ever be possible through traditional methods. And it is faster, too. "The 3D printer allowed us to rapidly prototype the part, find the best design and move it quickly to production," says Bill Millhaem, general manager for the GE90 and GE9X engine programs at GE Aviation. "We got the final design last October, started production, got it FAA certified in February, and will enter service next week. We could never do this using the traditional casting process, which is how the housing is typically made."

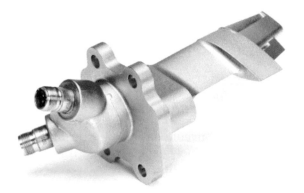

FIGURE 1.18
3D Printed T-25 sensor from GE aviation.

1.5.9 On-Demand Parts in Space

Parts of space exploration vehicles have been traditionally manufactured using standard manufacturing techniques here on earth. In many cases, when the part would fail in space, there was no way to replace the part unless the replacement part was flown into orbit. This is expensive and time-consuming. Recently, 3DP technology has been used to manufacture parts for the next generation space exploration vehicle. This is important in two ways: first, the maturity of 3DP technology is getting to the level where the parts manufactured have sufficient properties and quality to be used in space exploration applications (usually space exploration applications require the parts that are extremely complex and have very specific properties, and high-precision tolerances); second, with the development of 3D printer that could manufacture parts in microgravity of the space vehicle, it would be possible to manufacture replacement parts directly in orbit. The idea is to be able to manufacture entire space ships in orbit. NASA's next generation space exploration vehicle consists of approximately 70 3D printed parts; yet, they are developed on the ground here on earth, which elongates the supply chain drastically. Printing parts on demand directly in space would significantly reduce the cost and planning cycles required to send a rocket into space with necessary replacement and repair tools. 3DP on-demand parts in space is what being currently investigated by groups like Made in Space and Lunar Buildings. In collaboration with NASA, Made in Space is conducting zero-gravity tests to experiment 3DP on the International Space Station, which would allow astronauts to print tools and parts when required. Figure 1.19 shows the team from Made in Space conducting their 3D printer tests in microgravity of the Zero-G airplane that flies in parabolic flight path allowing for approximately 30–40 s of weightless periods.

FIGURE 1.19
Made in Space 3D printer tests conducted in the microgravity of zero-g airplane.

1.5.10 Medical Applications

3DP technology has been applied in medical field since the early 2000s. At first, 3DP technology was used to first manufacture dental implants and custom prosthetics. Since then, the medical applications of 3DP technology have evolved considerably. Recently published reviews describe 3DP to produce bones, ears, exoskeletons, windpipes, jaw bones, eyeglasses, cell cultures, stem cells, blood vessels, vascular networks, tissues, and organs, as well as novel dosage forms and drug delivery devices. Currently, 3DP applications in medical field can be categorized in the following categories: tissue and organ fabrication, prosthetics, implants, anatomical models, drug delivery, and dosage forms [9].

1.5.10.1 Bioprinting Tissues and Organs

Tissue or organ failure due to many factors is a critical medical problem. Current medical treatment for organ failure relies mostly on organ transplants. However, there is a large shortage of human organs available for transplants. Additional problem with organ transplantation is that it is difficult to find a good tissue match. Therapies based on tissue engineering and regenerative medicine are being pursued as a potential solution to the organ donor shortage. Although still very much in its infancy, 3D bioprinting offers important advantages beyond the traditional regenerative method. Organ printing takes advantage of 3DP technology to produce cells, biomaterials, and cell-laden biomaterials individually or in tandem, layer by layer, directly creating 3D tissue-like structures. Various materials are available to build the scaffolds, depending on the desired strength, porosity, and type of tissue, with hydrogels usually considered as the most suitable for producing soft tissues.

Although 3D bioprinting systems can be laser-based, inkjet-based, or extrusion-based, the inkjet-based bioprinting is most common. This method deposits "bioink," droplets of living cells or biomaterials, onto a substrate according to digital instructions to reproduce human tissues or organs. Multiple printheads can be used to deposit different cell types (organ-specific, blood vessel, and muscle cells), a necessary feature for fabricating whole heterocellular tissues and organs.

1.5.10.2 Implants and Prosthetics

Customized medical implants and prosthetics can be manufactured in nearly any imaginable geometry through the translation of X-ray, MRI, or CT scan data into an STL 3DP format that is ready for manufacturing on 3DP technology. In this way, 3DP has been used successfully in the health care sector to make both standard and complex customized prosthetic limbs and surgical implants, sometimes within 24 h. This approach has been used to fabricate dental, spinal, and hip implants. Previously, before implants could be used clinically, they had to be validated, which is extremely time-consuming. The ability to quickly produce custom implants and prostheses solves a clear and persistent problem in orthopedics, where standard implants are often not sufficient for some patients, particularly in complex cases. Previously, surgeons had to perform bone graft surgeries or use scalpels and drills to modify implants by shaving pieces of metal and plastic to a desired shape, size, and fit. This is also true in neurosurgery. Skulls have irregular shapes, so it is hard to standardize a cranial implant. In victims of head injury, where the bone is removed to give the brain room to swell, the cranial plate that is later fitted must be perfect. Although some plates are milled, more and more are created using 3D printer technology, which makes it much easier to customize the fit and design. There has been a lot of commercial and clinical success regarding the 3DP of prostheses and implants. A research team at the BIOMED Research Institute in Belgium successfully implanted the first 3D printed titanium mandibular prosthesis. The implant was made by using a laser to successively melt thin layers of titanium powders. 3DP has already had a transformative effect on hearing aid manufacturing. Today, 99% of hearing aids that fit into the ear are custom-made using 3DP techniques. Figure 1.20 shows the 3D printed hearing aid shell.

1.5.10.3 3D Printed Dosage Forms and Drug Delivery Devices

3DP technologies are already being used in pharmaceutical research and fabrication, and they promise to be transformative. Advantages of 3DP include

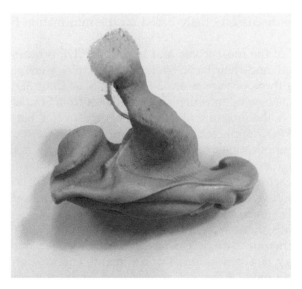

FIGURE 1.20
3D Printed hearing aid shell.

precise control of droplet size and dose, high reproducibility, and the ability to produce dosage forms with complex drug-release profiles. Complex drug manufacturing processes could also be standardized through the use of 3DP to make them simpler and more viable. 3DP technology could be very important in the development of personalized medicine too. The purpose of drug development should be to increase efficacy and decrease the risk of adverse reactions, a goal that can potentially be achieved through the application of 3DP to produce personalized medication.

1.6 Summary

In this chapter, the basic concepts of 3DP were introduced and discussed. 3DP refers to manufacturing of a physical, 3D model using AM methods. Some of the advantages of 3DP are reduced lead-time to bring a product to the market, improved part geometry, and early detection of any design problems. Many industries are using 3DP more and more in their manufacturing processes.

The basic 3DP process involves the creation of a CAD model of a part, converting the CAD model of the part to STL format, slicing the STL file into 2D cross-sections, 3DP the part, and then cleaning and finishing the part. Although the modern concept of 3DP was conceived and promoted in the 1980s, the origin of this technology dates back to the 1890s. General trends in the sales and applications of personal/office space and commercial 3DP machines were discussed briefly based on the information from Wohler's Report.

About three of the most iconic and important 3DP processes have been briefly reviewed, and their basic operating principles were studied in this chapter. These three processes are as follows: SLA from 3D systems, SLS from 3D systems fused deposition method (FDM) from Stratasys.

Lastly, some of the most interesting applications of 3DP have been briefly discussed. The early applications of 3DP were geared toward product development. Most applications can be subdivided into three main categories: prototype design evaluation, prototype for function verification, and prototype for manufacturing process validation. In addition, 3DP is being used in reverse engineering, casting and pattern making, RT, medical applications, aerospace applications, and many more areas.

3DP is a relatively new technology that is impacting the way that new products are manufactured. Despite some challenges and difficulties, 3DP technology is a continuous improvement. The idea of 3DP process will continue to move from 3DP to RM in the near future as more 3DP applications become apparent, and more advanced and faster machines become available.

1.7 Questions

1. What is a prototype? Why do you need a prototype? What are the potential uses for a prototype?
2. What are the considerations for using physical prototypes?
3. What is AM?
4. Define RP and 3DP.
5. What are the major advantages of RP and 3DP?
6. What is the difference between 3DP and conventional machining?
7. What are the main steps in making a prototype using 3DP?
8. Name three 3DP machine producers and describe the principles of each system.
9. Show the positive impact of 3DP on the product development cycle using flow diagrams.
10. Show the historical growth trend for both RP and 3DP.

References

1. R. Noorani, *Rapid Prototyping: Principles and Applications,* John Wiley and Sons, Englewood, NJ, 2006.
2. P. Jacobs, *Rapid Prototyping and Manufacturing & Fundamentals of Stereolithography,* SME Publications, Dearborn, MI, 1992.
3. T. Wohlers, Wohlers report 2016. 3D printing and additive manufacturing state of the industry. Annual Worldwide Progress Report.
4. C. Coward, *Idiot's Guide As Easy As It Gets 3D Printing,* Alpha by Penguin Group, New York, 2015.
5. C.C. Kai and L.K. Fai, *Rapid Prototyping: Principles and Applications in Manufacturing,* John Wiley & Sons, New York, 1997.
6. M. Frauenfelder, *Make: Ultimate Guide to 3D Printing,* Maker Media, Inc., 2014, 1–112.
7. G. Lin and L. Chen, An intelligent surface reconstruction approach for rapid prototyping manufacturing, *The Fourth International Conference on Control, Automation, Robotics and Vision,* Singapore, 43–47, December 3–6, 1996.
8. B. Fritz and R. Noorani, Form sheet metal with RP tooling, *Advanced Materials & Processes,* Vol. 155, No. 4, pp. 37–42, 1999.
9. I. Gibson, L.K. Cheung, S.P. Chow, W.L. Cheung, S.L. Beh, M. Savalani, and S.H. Lee. The use of rapid prototyping to assist medical applications, *Rapid Prototyping Journal,* Vol. 12, No. 1, pp. 53–58, 2006.

2

How Does 3D Printing Work?

2.1 Introduction

The objective of 3D printing is to be able to quickly and inexpensively fabricate any complex-shaped, 3D part directly from CAD data without any usage of tooling. 3D printing is an example of an additive manufacturing process. In 3D printing, a solid CAD model is sliced into layers of predetermined thickness by a special slicing software that is proprietary for each 3D printing machine. These sliced sections define the overall shape and geometry of the part collectively when stacked on top of each other.

The focus of this chapter is on those system elements that affect the shape of the part: the CAD file, the Stereolithography (STL) file, problems and repairs of STL files, and other file formats, which could be used instead of STL file. In addition, the general principles of 3D printing technology will also be discussed in this chapter [1].

2.2 3D Printing and Conventional Manufacturing

3D printing is essentially a part of automated fabrication process, a technology that allows to create 3D physical parts from their digital (CAD) designs [2].

Automated fabrication has a few advantages over typical manual fabrication and molding processes. Some of these advantages are its ability to use computer-aided design, quick design changes possible with edits of CAD model, and precise dimensioning.

All fabrication processes, manual or automated, can be classified as subtractive or additive. An example of subtractive process is the typical computer numerical control (CNC) machining and of additive process is 3D printing, such as fused deposition modeling (FDM), stereolithography apparatus (SLA), and selective laser sintering (SLS). These processes are shown in Figure 2.1.

Subtractive manufacturing—a manufacturing process where a solid block of material larger than the final size of the finished part is taken, and then the material is removed until the desired shape of the part is obtained. A complex process that requires 5 axes numerical control (NC) for parts with complicated geometries.

Subtractive processes include forms of machining processes such as computer numerical control (CNC). The most widely used examples include milling, turning, drilling, planning, sawing, grinding, electrical discharge machining, laser cutting, water-jet cutting, and many other methods.

Additive manufacturing—a manufacturing process that involves manipulation of very small quantities of material at a time so that successive pieces of it combine in the right form in order to produce the desired part directly from the CAD model. Very simple process, which requires manipulation of "2D" layers for most complex parts. For example, all of the 3D printing processes, such as FDM, SLA, SLS, laminated object modeling (LOM), and 3D printing,

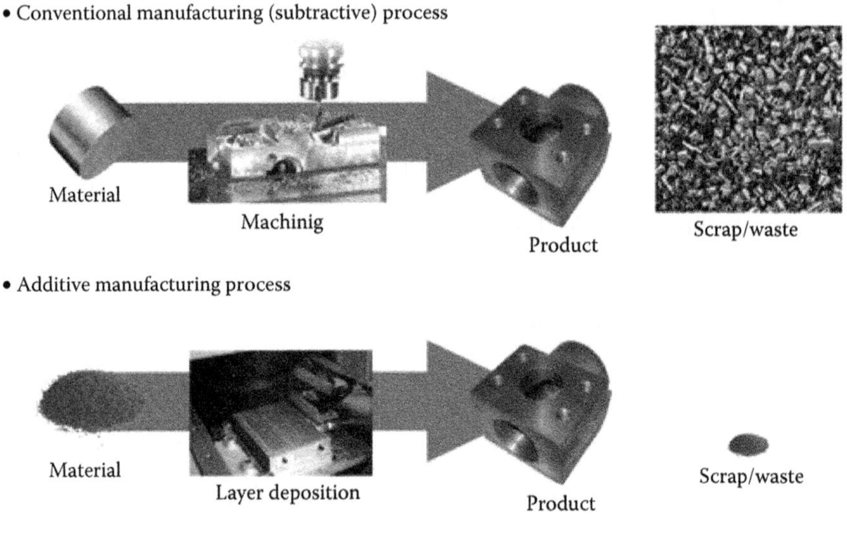

- Conventional manufacturing (subtractive) process

Material   Machinig   Product   Scrap/waste

- Additive manufacturing process

Material   Layer deposition   Product   Scrap/waste

FIGURE 2.1
Subtractive vs. additive manufacturing.

fall into the additive manufacturing category because the parts manufactured using 3D printing process are made one layer at a time from bottom up.

Another important difference between additive and subtractive manufacturing is the complexity of the parts that can be produced with each process in a limited time. For example, if a part is to be produced has a very complex geometry such as the part shown in Figure 2.2 with many geometrical features, a typical subtractive process such as 3-axis CNC machining will not be able to create all the features. More complex 5 or more axes CNC machine needs to be used in conjunction with another specialized machine and hand tool to manufacture such part. One can already see that this process uses a number of different machines and is also time-consuming. In contrast, 3D printing technology could manufacture the part with all its features using a single machine in relatively short time, and the process is completely automated.

Benefits of additive manufacturing are the creation of lightweight and highly customizable parts, which can be produced without complex machinery. This also means that the unique consumer goods can be easily created and that parts of an assembly can be consolidated into fewer items, making it much easier to complete the assembly.

Major differences between additive and subtractive manufacturing are summarized in Table 2.1.

Although additive manufacturing has been around for a while, there are still some challenges that need to be overcome: system reliability, need for closed-loop controls, expense, need of larger build volumes, build speed, etc.

It is important to note that two or more subtractive and additive manufacturing processes can be combined to form a *hybrid* process. Hybrid processes

FIGURE 2.2
Typical impeller.

TABLE 2.1

Differences between Additive and Subtractive Manufacturing

Additive Manufacturing	Subtractive Manufacturing
Part consolidation, entire assembly manufactured	Many different parts for the same assembly have to be manufactured individually
Still relatively expensive	Cheap due to mainstream use and availability
Special materials	Standard materials
Typically not the best accuracy	High accuracy is easily achievable
In many cases, surface finish is not optimal	Excellent surface finish
Maturing reliability of machines	Generally, very reliable machines
Part size is a limitation	No part size limitations
Can create thin walls	Thin walls are hard to create
Unlimited length of small holes conforming cooling channels	Difficult to create small holes and features
Machines run autonomously	Machines need constant attention
More complex part is more cost saving is achieved	Complex parts are very expensive
Automated preprocessing	Preprocessing must be done manually

are expected to contribute significantly to the production of goods in the future. Progressive press working is an example of hybrid machines that combine two or more fabrication processes. In progressive press working, a hybrid of subtractive (as in blanking or punching) and formative (as in bending and forming) is used.

2.3 Basics of 3D Printing Process

All parts that are manufactured with both the current and evolving 3D printing processes have several procedural features that are carried out during manufacturing process in common. A solid or surface CAD model of the part to be produced must first be created, and then it is electronically sectioned into layers of predetermined thickness with special slicing software that is proprietary to each 3D printing machine. These layered cross sections define the shape of the part collectively. Information about each sliced section is then electronically transmitted to the 3D printing machine layer by layer. The 3D printing machine processes materials only at "solid" areas of the section. Subsequent layers are sequentially processed until the part is complete. It is this sequential, layered, or lithographic approach to parts manufacturing that defines 3D printing. The 3D printing process basically uses the following five basic steps to manufacture 3D physical parts [3]:

1. Creation of a CAD model of the design.
2. Conversion of the CAD model to an STL file format.
3. Slicing the STL file into 2D cross-sectional layers.
4. Growing of the prototype.
5. Postprocessing.

This five-step process is shown in Figure 2.3.

2.3.1 Creation of Solid Model

The first step in creating a 3D printed part is the creation of a CAD model of the part. 3D printing process requires the part to be of fully closed geometry and watertight such that even if one were to pour water into the volume of the model, water would not leak.

A solid is a volume that is completely bounded by surfaces, which means the edges of all surfaces must be coincident with one, and only one other surface edge. Unlike wireframe and surface modeling, solid modeling stores volume information. A CAD solid model not only captures the complete

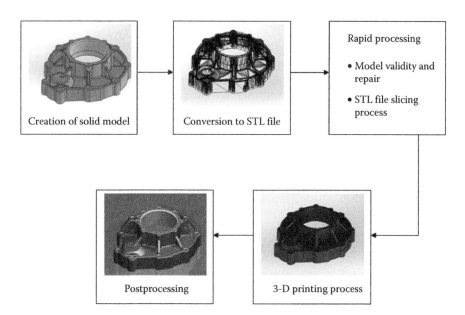

FIGURE 2.3
Five-step process of 3D printing.

geometry of an object, but also can differentiate the inside and the outside of the space of that object. Many other volume-related data can be obtained from the model if needed. Solid models can be created using any CAD software package, such as AutoCAD, PTC Creo, CATIA, or Solid\Works, or many other commercially available solid modeling programs. Figure 2.4 shows a typical CAD model view of a mechanical part in CAD software, in this case SolidWorks.

2.3.2 Conversion to STL File

Once a CAD model is created and saved, it is then converted to a special file format known as STL format. This file format has originated from 3D Systems, a company that pioneered the STL process. Actually, the Albert Consulting Group under contract to 3D Systems developed the STL file format to support the new revolutionary manufacturing technology called stereolithography apparatus (SLA). Though not ideal, it is sufficient to meet the needs of today's 3D printing technology, which generally manufactures mono-material parts. The success of this file format has been impressive. Today, a decade later, the STL file format remains the *de facto* standard for the 3D printing industry.

The success of the STL file format is due to its sufficiency, simplicity, and monopoly. Its mathematical sufficiency stems from the fact that it describes a solid object using a boundary representation (B-Rep) technique. An STL file format represents the object modeled in the CAD software as a series of triangles. These triangles together as a whole are used to approximate the surface of the object. STL files are used because most prototyping machines can follow the linearity of triangles much easier than any other object while still maintaining the object's integrity. In short, an STL file is nothing more

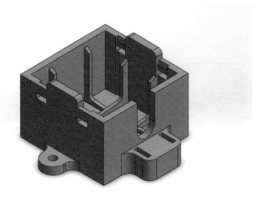

FIGURE 2.4
CAD model of a mechanical part created in SolidWorks.

than a list of x, y, and z coordinates triplets that describe a connected set of triangles to approximate the surface of an object.

Most CAD/CAM software vendors supply the STL file interface. Since 1990, many CAD/CAM vendors, such as SolidWorks or AutoCAD, have developed and integrated this interface into their systems. Tessellation is the process of approximating a surface using triangles. The CAD STL file interface performs surface tessellation and then outputs the information to either a Binary or ASCII STL file format. An example of fine tolerance and ASCII output STL model approximation of the part is shown in Figure 2.5.

As stated earlier, the STL files can be expressed in either Binary or ASCII format. Both styles essentially complete the same task (converting a model into a format that is able to be read and manufactured), but there are differences between the two formats, as noted in Table 2.2.

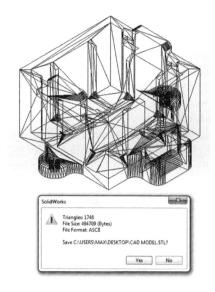

FIGURE 2.5
STL file approximation of the mechanical part geometry.

TABLE 2.2

STL File Output Types: Binary vs. ASCII

Binary	ASCII
Computerized language, able to be read by a computer	Default saving format for most software, can be read and understood by a human
More compact and efficient, easier to move through network/transmit	Not very efficient, not recommended if moving files through a network
Quicker processing speed, smaller file type	Slower to process, larger file sizes

There are three basic steps followed when creating an STL file: the selection of the part to be converted, choosing the tolerance for the conversion process, and the approximation of the object by triangles that is then saved as an output file.

As mentioned previously, both the surface and solid parts of a model can be converted into the STL format, but it is more difficult for surface models to be created. When a surface model is being processed, there are more steps to follow to ensure that the object is properly produced. First, all places where different surfaces meet must be determined. When these surfaces are changed into a series of triangles, it is important to ensure that all edge vertices line up and match to avoid zero-thickness points (places where the object cannot be manufactured). After this, it is important to check that the normal for the object (the orientation for a surface) points to the outside of the model. Once this is checked, the rest of the object can be converted into triangles and saved as an output file. Many times, this process can be completed by the CAD software, but it is still important to do a quick recheck of all these steps.

For both a regular model and a surface model, there are also a series of interface tolerance and appearance options that must also be addressed. These tolerances include the triangulation tolerance, adjacency tolerance, and auto-normal generation as well as the appearance options of triangle display and header information.

Triangulation tolerance is the tolerance that determines how smooth the approximation of the object will be. A smaller (referred to as tighter) triangulation tolerance will produce more triangles that lie along the edges and approximate the surface better. However, this will increase the amount of time it takes to manufacture the object. The default for most triangulation tolerances is 0.0025 inches or 0.05 mm, which creates a relatively accurate object without increasing the manufacturing time too much. The effects of creating a tighter triangulation tolerance are shown through the triangular approximation of a sphere in Figure 2.6.

In addition to the triangulation tolerance, the adjacency tolerance also affects the creation of the STL file. The adjacency tolerance does not affect the processing of solids, but instead is used to determine how closely two

FIGURE 2.6
Effects of tightening the triangulation tolerance.

surfaces will be attached to one another. With the use of triangles in STL files, it is not possible to create perfectly curved surfaces so there are small edges between every surface. Adjusting the adjacency tolerance will affect how close these edges are to each other and how smooth an approximated curve will be. The default value for the adjacency tolerance is 0.005 inches or 0.12 mm. Again, tightening the adjacency tolerance will increase the production time, but will make a more naturally curved surface.

The final tolerance for STL files is the auto-normal generation. Normals are important as they determine the orientation of a surface relative to a predetermined axis, which is often chosen by the CAD software. Most CAD software have auto-normal generation, meaning that the normal for every surface is chosen automatically, saving the designer of the object time. Auto-normal generation typically will produce normal pointing to the outside of the object, which is necessary to print a surface model. This is done by choosing a base surface (generally the first surface created in the CAD software by the creator), checking the normal of that surface, and then calculating all other normals from this surface. The default for auto-normal generation is typically on, but it can be turned off if the designer wants to manually adjust normals to their choosing.

Following these tolerances, there are two appearance options that can be adjusted. The triangle display option can be turned on and off, which determines if the triangles are shown in when the object is saved as in the STL format. In addition, the header display can be toggled on and off, which determines if the text notes on the object are shown or not.

2.3.3 Slicing the File

The next step in the process is to slice the STL file. In this step, STL file is sliced, meaning that the triangles are prepared to be manufactured. Each manufacturing device uses its own software to slice a file. In order to slice the STL file, first STL file slicing software must be opened and the desired STL file is imported. Once it is imported, many different options for slicing become available, such as the layer thickness, quality of print, extrusion temperature, material, in fill percentage, or the option to add support material or print rafts. These options are selected based on the design needs. Most of these options are found in the "settings" tab in the manufacturing machine's software. An example of this is shown in Figure 2.7, which shows the software for the MakerBot Replicator (a 3D printer) with the layer thickness, quality, extrusion temperature, material, infill percentage, and support material/rafts circled in red, light green, dark green, yellow, orange, and black, respectively.

Once the slicing options are selected, STL file is then automatically sliced and prepared for fabrication. The more complex the object, the more triangles are required, and thus, the bigger the file that makes up the CAD model as well as a support structure for the part to be produced properly.

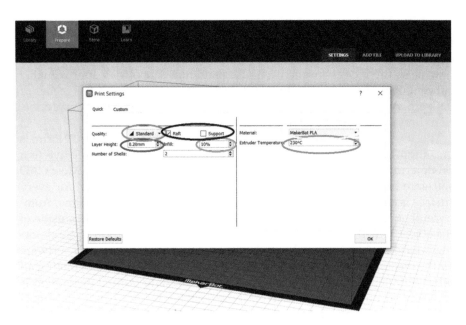

FIGURE 2.7
"Settings" menu with slicing options.

2.3.4 Making the Prototype

Once the STL file of the part to be produced is sliced, it is loaded into the 3D printer. The appearance of a file loader into MakerBot Desktop, the software used for the MakerBot Replicator 3D printer, is shown in Figure 2.8. Once the object is loaded into the 3D printer, the machine runs autonomously until the part is complete. 3D printing machines build one layer (slice) of the part at a time from materials such as polymers, paper, or powdered metals. Most machines are fairly autonomous needing little human intervention. Build times vary depending on size, number of parts required, and machine settings used for the build. For the part shown in Figure 2.8, it would take approximately 9–11 hours to build it using MakerBot Replicator 3D printer with settings that yield the best part quality.

2.3.5 Postprocessing

The final step in the manufacturing process is postprocessing, which is removing excess materials and cleaning the part. For different production styles, this also sometimes requires curing the object, as is the case with SLA or SLS production styles. Often this step requires special knowledge or actions to ensure that the part is not damaged. Otherwise, the part may be damaged and may need to be manufactured again.

The tasks for postprocessing for different rapid prototyping (RP) processes are given in Table 2.3, while Table 2.4 summarizes the necessary

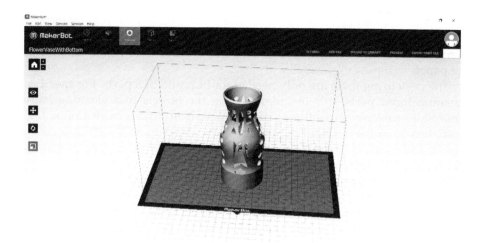

FIGURE 2.8
Layout of part in MakerBot Desktop.

postprocessing tasks for the MakerBot Replicator and Formlabs 1+, two 3D printers at Loyola Marymount University.

For the different types of RP systems and 3D printers, the cleaning, post-curing, and finishing varies greatly. The cleaning for FDM machines and the MakerBot Replicator includes the removal of support materials and excess filament through sanding the objects. For SLA machines and the Form1+,

TABLE 2.3

Postprocessing Tasks for Various Rapid Prototyping Systems

Postprocessing Tasks	Fused Deposition Modeling (FDM)	Stereolithography Apparatus (SLA)	Laminated Object Manufacturing (LOM)	Selective Laser Sintering (SLS)
Cleaning	X	X	X	X
Postcuring		X		X
Finishing	X	X	X	X

Note: X means it is applicable.

TABLE 2.4

Postprocessing Tasks for the MakerBot Replicator and Formlabs 1+

Postprocessing Tasks	MakerBot Replicator (FDM Style)	Formlabs 1+ (SLA Style)
Cleaning	X	X
Postcuring		X
Finishing	X (optional)	X

Note: X means it is applicable.

this includes cleaning off excess resin and pulling off any support materials. For SLS machines, the cleaning includes removing leftover powder from the process, while LOM cleaning entails the removal of excess wood-like blocks of paper.

The postcuring tasks are only needed for SLA and SLS parts. For the SLA manufactured parts, this includes putting the objects in a cleaning bath (normally rubbing alcohol) and then leaving it in a place with minor UV radiation to finish curing. For SLS parts, the object must be left in a clean environment where the part can set and solidify after the bonding materials for the powder finishes solidifying.

Part finishing is done once the part is postcured or cleaned. Part finishing can include another round of sanding or removing any excess material carefully or adding a protective covering to the object to keep it from being damaged. For parts that are printed on an FDM or SLA machine, it is easy to paint the objects using acrylic paint to get a desired appearance from the objects.

2.4 Problems with the STL File Format

The STL format is the most popular and common file type in RP and 3D printing as it meets all the standards of these types of manufacturing. However, the STL format still has some inadequacies. These problems often stem from the fact that the use of triangles to approximate curves means there can be edges or gaps as well as differences in tolerance usage or normals [4].

Gaps are the most common issue with the STL format because as a solid model is converted into an STL format, the model's curves are replaced with a simplified geometry (triangles), meaning small fluctuations in the curves can create minor geometric anomalies, thus making holes or gaps in the surface. Problems with gaps typically arise when an STL file is approximating surfaces with large curvature. An example of a problem with gaps is shown in Figure 2.9a.

Another issue that can arise from STL file formats is inconsistent or incorrect normals. Occasionally, objects can have normals that are flipped and not pointed outward as they should be. This can result in divots or dents in an object that are not desired. This problem normally occurs only when creating a highly complex model or when auto-normal generation is turned off. Figure 2.9b shows an example of an object within consistent and flawed normals.

A final issue that can developed when using STL file formats have issues within consistent tolerances. Sometimes, two STL files are combined to create a prototype, and if these two files have been created using different tolerance values, it will lead to inconsistencies similar to the other two issues of gaps or incorrect normals.

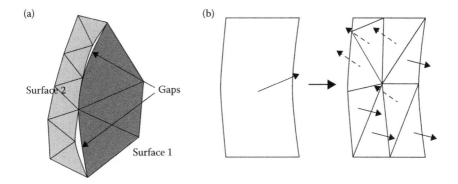

FIGURE 2.9
(a) Polygonal approximations resulting in gaps and (b) polygonal approximations resulting in inconsistent and incorrect normals.

2.5 Other Translators

Since there are problems and inefficiencies associated with the STL file format, a lot of work has been done in search. Some alternative translators that are new and becoming more common are virtual reality modeling language (VRML), additive manufacturing format (AMF), 3D manufacturing format (3MF), and object file (OBJ) formats. In addition, some older alternative translators that are still in use are 3D studio (3DS), initial graphics exchange specification (IGES), Hewlett-Packard graphic language (HPGL), and computerized tomography (CT data).

2.5.1 Modern File Formats

2.5.1.1 VRML File

The VRML is a newer digital 3D file type that can include color, which makes it advantageous to use for dual extruder printers that can print in more than one color. VRML is becoming more common among 3D printing enthusiasts, but the interface is slightly more complex than STL, so it has not become a commonplace file type. Also, the manufacturability of the objects from the VRML type is not of high quality as other types.

2.5.1.2 AMF File

The AMF is an extendable markup language (XML) file format that also contains support for color. In addition, it is a compressed file type that takes up half the size of an STL file. AMF is not widely used, but its support for color and smaller file type is convenient for long-term use. However, the AMF type has less features than an STL file type, which does present a significant flaw.

2.5.1.3 3MF File

The 3MF file format is a new format that is coming from Microsoft. The 3MF is another format that features color and a smaller file type without the cost of features that the STL file type has. The 3MF also is an easy file type to transfer and puts out quality manufacturability in products. The file type stems from the 3MF consortium made up of companies, such as Microsoft, HP, Autodesk, and other companies, that are getting involved in 3D printing. The 3MF is not widely used yet, but is seen as the file type of the future and as a contender to replace STL format in the long run.

2.5.1.4 OBJ File

The final modern day file type is the OBJ type, which is a format that is becoming popular in 3D printing for its ease of use. The file type presents only the 3D geometry with labeled attributes such as vertex normals and geometric vertices, making it easy to understand the dimensions of the object and find flaws easily. The downside to the OBJ file format is that it does not create the highest quality of prints as the triangulation process takes less precedence in OBJ files.

2.5.2 Older File Formats Still in Use

2.5.2.1 3DS

The 3DS file type was one of the formats developed for usage in the Autodesk 3D studio software package. It was popular in the mid-1990s due to its relatively simple interface and ease of usage. The 3DS uses a binary file format based on blocks that contain print data developed in a hierarchical structure. This means that it is quite quick to slice and develop files. However, the blocks that are used are not as accurate as triangles in other formats, meaning that the prints that have curves are extremely hard to create.

2.5.2.2 IGES File

IGES is a common format to exchange graphics information between various CAD systems. It was initially developed and promoted by the then American National Standards Institute in 1981. The IGES file can precisely represent both geometry and topological information for a CAD model, thus making it more accurate than the STL file format. An IGES file contains information about surface modeling, constructive solid geometry, and boundary representation (B-rep). It can precisely represent a CAD model by providing entities of points, lines, arcs, and splines as well as surface and solid elements. The primary advantage of IGES format is its widespread adoption and comprehensive coverage. However, there are some disadvantages associated with the IGES format as it relates to its use as a 3D printing format as it includes redundant information for 3D printing systems, its algorithms for slicing are far more complex than STL files, and the support structures cannot be created in IGES formats. IGES is a very good interface standard for exchanging information between various CAD systems. It does, however, fall short of meeting the standards for 3D printing systems.

2.5.2.3 HPGL File

HPGL has been the standard data format for graphic plotters for many years [1,2]. Generally, data types are 2D representing lines, circles, splines, text, etc. Many commercial CAD systems have the interface to output HPGL format, which is a 2D geometry data format and does not require slicing. However, there are two major problems with the HPGL format. First, since HPGL is a 2D data format, it leaves hundreds of small files needing logical names and transformation. Second, all the required support structures must be generated in the CAD system and sliced in the same way.

2.5.2.4 CT Data

CT scan data is a new format that is primarily used in medical imaging. This format has not been standardized yet. Various versions of CT data formats are proprietary and vary from machine to machine. The CT scan generates data as a grid of 3D points, where each point has a varying shade of gray indicating the density of body tissue present at that point. Data from CT scan is being regularly used to prototype skull, femur, knee, and other biomedical components on FDM, SLA, and other 3D printing systems. The CT data essentially consists of raster images of the physical objects being imaged.

Models using CT scan images can be made using CAD systems, STL interfacing and direct interfacing. CT data is used to make human parts like leg prostheses, which are used by doctors for implants. However, the main problem with CT image data is the complexity of the data, and the need for a special interpreter to properly process this data.

2.6 Future Manufacturing Format Developments

Most RP systems make mono-material parts, although FDM, solid ground curing (SGC) and a few other 3D printing systems use secondary material for support structures. For these systems, the STL file format is sufficient as it describes the location of the build and support materials. Future 3D printing systems will be challenged to increase the range of automated fabrication capabilities where a few different materials will be used. In his writings, Bohn contemplates that the interfacing advances listed below might be expected to emerge and mature to the commercial market over the next several years.

Another possible improvement for future formats is internal coloring in prototypes. Materialise in Leuven, Belgium has demonstrated the feasibility of selectively coloring parts prototyped by the STL process. For this purpose, a special resin is used that changes its color depending on the local curing condition.

In addition to this, changing materials in a single print is another development that could greatly improve the prototyping process. Fitzgerald, another prototyping entrepreneur, has demonstrated that prototype parts can be fabricated with a continuously changing blend of materials using an FDM-like process. Fitzgerald and Bohn have also investigated graded materials for powder-based (SLS) systems. This technique will allow engineers to design parts with locally defined mechanical properties. Similar research emphasizing variation of 3D printing material composition under software and computer control is also being carried out in companies and universities including MIT and UT Austin.

A final possible improvement for prototyping was developed by Prinz et al. They have demonstrated that foreign components such as components that are more efficiently fabricated separately by other means can be imbedded into composite products.

It is widely believed that the future will bring exciting new capabilities to the world of RP. New opportunities will empower engineers to design complex parts that they could only dream of building a few years ago. As the power of computer control increases, so does the power and flexibility of RP machines.

2.7 Case Study: Design and Printing of Eye Bracket

This section describes a complete design and 3D printing process of a simple eye bracket part. Step-by-step procedure for making the prototype as well as the hardware and software used for making the prototype will be thoroughly discussed in this case study. After the part was created, dimensional verification was carried out with respect to nominal dimensions that were specified in engineering drawing.

2.7.1 Introduction

The eye bracket is a very common engineering part that is used to make parts in a simple and convenient way using standard fasteners, and it allows for easy future disassembly if needed. A 3D CAD file of eye bracket was created using SolidWorks software, and it was 3D printed using MakerBot 3D printer. MakerBot uses FDM technology to build the parts in a relatively short period of time and in a very inexpensive way. The material used in this case study was polylactic acid (PLA) plastic filament. It is heated above its melting point and then extruded through a nozzle onto a build platform. When one cross-sectional layer of the part is finished, the platform is lowered and a new cross-sectional layer of the part is deposited on top of the previous layer. FDM process is an example of additive manufacturing technology. This technology offers greater capability and flexibility in terms of ability to produce complex geometry parts compared to the traditional subtractive manufacturing techniques such as CNC machining.

Using engineering drawing of the eye bracket and all the dimensions and features that are contained in the drawing, a 3D CAD model of the eye bracket was created in SolidWorks CAD software. Figure 2.10 shows a 2D drawing of the eye bracket part. The advantage of creating a model of the part in SolidWorks is that the dimensions can be easily changed to allow for easy and quick design changes if needed. This software allows almost any geometry to be created and then saved as an STL file.

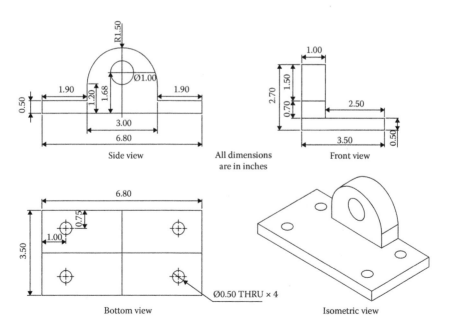

FIGURE 2.10
2D engineering drawing of eye bracket.

2.7.2 Project Procedure

The 3D printing process for this engineering prototype consisted of five major steps: CAD modeling process; conversion of CAD model file to STL file; slicing the STL file; 3D printer setup, actual 3D building process, and postprocessing; evaluation and verification of the produced prototype. An image of MakerBot 3D printer used in this case study can be seen in Figure 2.11.

1. CAD modeling process
   a. A detailed engineering drawing of the part was obtained.
   b. Using the engineering drawing, a 3D CAD model was created in SolidWorks CAD software.
   c. If there is more than one component to the model that is to be 3D printed, the part geometry must be perfectly mated without any gaps. This is because the model should be one solid object in order to convert later to an STL file.
   d. The native SolidWorks CAD file format was then converted into an STL file by choosing *File-Save As-STL file* commands.
   e. An STL file that was just created could be readily opened in MakerWare STL file slicing software.
2. STL file preparation
   a. An STL file that was created in the previous step is now ready to be loaded into MakerBot preprocessing software called

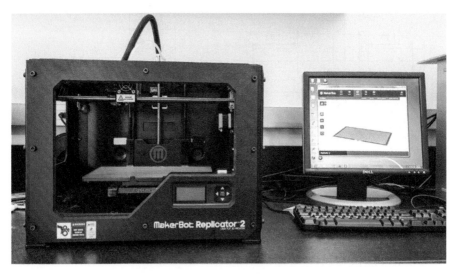

FIGURE 2.11
MakerBot 3D printer and supporting computer workstation.

MakerWare. Simply choosing *File*-Open commands and locating the desired file would load the part in STL format into the main window of MakerWare program.

b. Loaded part can now be visualized in 1:1 scale. Working volume of the MakerBot as well as the build plate is shown. Simple operations such as orientation of the part, positioning on the build table, and scale must be selected.

c. Orientation feature allows rotating the part around *x*, *y*, and *z*-axes in order to select the best suitable orientation for the build. This is achieved by clicking the *Turn* icon on the left-hand side of the MakerWare window on the computer screen. As an operator gains experience with the machine, the best possible orientation of the part for the build would be more intuitive to understand.

d. The next step is positioning. In this step, the part is positioned on the virtual build plate that would correlate to the location of the part on the build plate when it is completely built. Correct positioning of the part depends on many variables and the part geometry. For this case, simply having the part built in the middle of the platform will work as the part has relatively simple geometry. A position command is achieved by clicking the *Move* icon. It is important to mention that the bottom geometry of the part has to be coincident with the platform.

e. Once the orientation and positioning of the part have been performed, the last parameter to set left is scaling. This allows the make part larger or smaller than the original size. For the purposes of this case study, a scale of 1:1 was chosen as this part would be used for dimensional verification later.

f. Once all possible parameters and settings have been defined, simply clicking the *Print* button in MakerWare software would carry out proprietary MakerBot slicing algorithm automatically in the background without any additional user input, verify the slicing, and send the sliced STL file directly to MakerBot 3D printer for printing. Figure 2.12 shows a screenshot of the MakerWare software and the STL file of the eye bracket part just before the print command was sent to MakerBot 3D printer.

3. MakerBot set up and build

a. Before the printing process can be started, it is important to set up a MakerBot 3D printer, so that the built part would be as perfect as possible.

b. First and foremost, the build platform was leveled so that the build platform surface is perfectly horizontal.

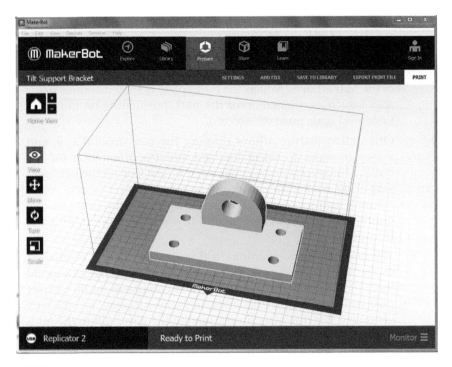

FIGURE 2.12
Eye bracket STL file in MakerWare.

  c. Once the leveling was completed, build platform was taken off and blue painters tape was applied to the build side. This helps in part removal after it is built.

  d. After build surface of the build platform was completely covered with blue tape, it was put back in place, and 3D printer was turned on and in a few minutes its extruding head reached 230°C temperature, which is the ideal temperature for 3D printing with PLA plastic material.

  e. As MakerBot extruding head stabilized at 230°C temperature, the build was started by simply pressing the *Print* icon in MakerWare software. Estimated time to build the part with 10% in fill by software was approximately 7 hours.

 4. Postprocessing

  a. After the 3D printing process was completed, the build platform was removed from MakerBot.

b. Support removal was accomplished with the use of simple hand tools (e.g., spatula, knife).

c. Sandpaper was also used to completely remove the base material from the base, and the part was ready for use. Figure 2.13 shows the finished eye bracket part.

5. Conclusions

This case study familiarizes the reader about the basics of 3D printing of simple parts. A simple engineering part was designed and manufactured using 3D printing technology. An image of the final part produced can be seen in Figure 2.13. A project like this can provide an invaluable experience to a group of young engineers in order to understand the process of making a design, and having a prototype to test before full production of the part begins. This is important when entering the industry, where knowledge of cutting-edge technology such as 3D printing can give young engineers an advantage over others.

This experience can also help prove the efficiency and flexibility of a CAM process, which involves 3D printing. A typical 3D printer can produce over eight different parts with completely different geometries in less than a week. This is actually remarkable for a machine that takes up no more room than a standard office laser printer. There are size and material limitations, but the advantages definitely outweigh the disadvantages, which are limited part size and less than optimum surface finish.

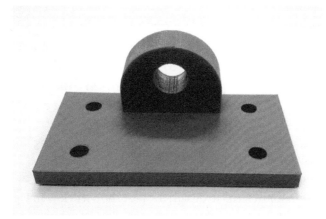

FIGURE 2.13
3D printed eye bracket part.

2.8 Summary

In this chapter, the basic concepts of 3D printing process have been introduced and discussed briefly. 3D printing is essentially a part of automated fabrication technology that allows to manufacture 3D parts directly from their numerical and digital design. There are two categories of manufacturing processes: additive and subtractive manufacturing. In many engineering applications, these two categories of manufacturing are used to create a hybrid manufacturing process. 3D printing process is an additive process. All 3D printing processes in general consist of five major steps: creation of solid CAD model using CAD software, conversion of solid CAD model to STL file format, slicing the file using a proprietary software package that is included with 3D printing machine, actual build phase of the part, and finally the postprocessing the part if needed. The most common 3D printing file format that is used in the industry and academia is the STL format, developed by the Albert Consulting Group for 3D Systems. Although the file format is not perfect, it is still being used as the *de facto* standard for RP industry.

2.9 Questions and Problems

1. What is a 3D printing system? Describe briefly a system showing all its components.
2. What are the purposes of data creation?
3. What is the most common file format used for 3D printing?
4. What are the advantages of the STL file format?
5. What are the shortcomings (problems) of the STL file format?
6. Mention several translators that can be used in place of STL format. What are some drawbacks or limitations of these?
7. What is postprocessing? Do you need postprocessing for all 3D printing systems?
8. Laboratory project: design and prototype a product
   a. Use a standard CAD to design the part shown below in Figure 2.14.
   b. Convert the drawing file into an STL file using conversion rules described in the text.
   c. Use an available 3D printing system (FDM, SLA, MakerBot Replicator2, FormLab1+ or any other 3D Printer) to prototype the part.

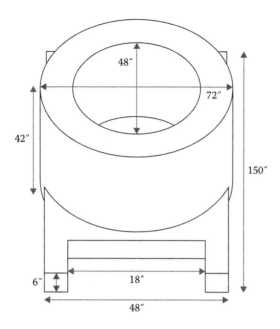

FIGURE 2.14

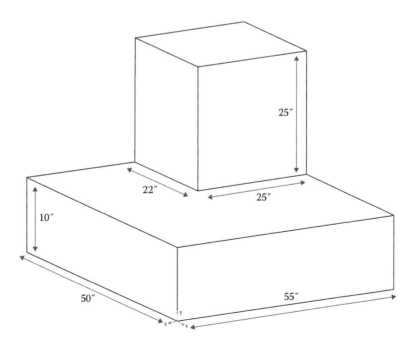

FIGURE 2.15

    d. Use postprocessing techniques to complete the prototyping cycle.

9. For the object shown below in Figure 2.15:

    a. Draw the part using SolidWorks or a similar CAD/CAM system

    b. Create STL binary file with chord height of 1.0 and 0.1.

    c. Discuss the changes that take place when chord height is varied (file size, number of triangles, etc.).

References

1. R. Noorani, *Rapid Prototyping-Principles and Applications*, John Wiley and Sons, Hoboken, NJ, 2006.
2. M. Burns, *Automated Fabrication; Improving Production in Manufacturing*, Prentice Hall, Englewood Cliffs, NJ, 1993.
3. P. Jacobs, *Rapid Prototyping and Manufacturing & Fundamentals of Stereolithography*, SME Publications, Dearborn, MI, 1992.
4. H. Bohn, File format requirements for the rapid prototyping technologies of tomorrow. *International Conference on Manufacturing Automation Proceedings*, Hong Kong, 1997.

3

Design of a 3D Printer

3.1 Introduction

The goal of this chapter is to design and fabricate a 3D printer that will be the most accurate and cost-effective printer possible, while still being faster than the rapid prototyping (RP) Lab's Stratasys FDM-1650 at Loyola Marymount University (LMU). This project is done in order to greatly increase the capability of the RP lab in addition to exposing students to the design and fabrication of a 3D printer. 3D Printing is a technology that takes information from CAD and builds 3D models using highly advanced, automated processes. The technology was developed in 1989 (still known as rapid prototyping) and since then has been expanding exponentially [1]. 3D printers allow for reduced lead times, production costs, and waste and improve the visualization of concepts. RP has been revolutionizing the manufacturing industry for a while. In order to complete the construction of the printer, the design is originally established in Solidworks to ensure the components would work together effectively. The frame is then constructed using V-slot aluminum and RP parts that are made in-house. Following this, the motors and extruders are assembled according to the design plans and the circuitry is wired through an Arduino-based programming system. Further steps needed to be taken were reconstructing the Y-axis motion and calibrating all aspects of the machine to finalize the assembly of the printer. The printer was completed in December 2014 and has been in operation since January 2015.

3.2 Necessary Parts

There is a multitude of necessary parts to make a 3D printer [2]. First, there are the extruders. Extruders are the parts through which the printing material is deposited. Within a fused deposition modeling (FDM) printer, an extruder consists of a nozzle, a hot end, and some form of mechanical drive to push the material through the nozzle. The hot end is a piece of metal that heats up to melt the thermoplastic into a malleable form before it is placed. The extruders need to be able to move in the X-, Y-, and Z-planes to be able to print a part properly. There are multiple different ways to do this, but typically, an FDM printer will consist of a combination of motors, screws, and belts. Motors drive both the screws and the belts to provide motion in each direction. Screws are typically used for the Z motion due to the fact that they provide very fine control over the layer thickness. In the X- and Y-planes, belts are used to reduce the complexity and size at the cost of some accuracy. All of these components are held within some sort of ridged frame.

3.3 Functional Description and Design Analysis

In order to fabricate this device, many materials and processes were utilized. Most of the parts used for this project were ordered; however, many of the complex and/or custom parts were 3D printed using the MakerBot 2 RP machine.

The fabricated printer can print nearly every RP material including acrylonitrile butadiene styrene (ABS) and polylactic acid (PLA), giving it a huge advantage over the other printers in the LMU RP lab. This is largely due to the hot end of the extruders. The hot end can print different materials due to the range of temperatures they can achieve. Each hot end in its current state can reach 300°C, and with few modifications as high as 400°C allowing them to print plastics with extremely high melting points. Additionally, the RP machine is designed for dual extruders, allowing it to print two different materials at once. This allows the printer to extrude either two different colors or both build and support material at the same time.

Three frame designs were considered for the 3D printer [3–4]. First, there was a square frame. A square frame design is more typical within the RP world, and this is because it is easier to work on the extra space between frame rails. This allows for easier servicing and assembly. However, this design is not as solid as other options out there and, therefore, can cause slight issues with accuracy. The second design considered was a delta frame. The delta frame provides a small footprint with a tall build volume and can be very useful for creating tall parts. In addition to this, the delta frames are cheaper due to the smaller amount of material necessary to build one and their reduced complexity in terms of the number of parts. On the other hand, though the build volume is very tall, it is not very large in the X- and Y-planes due to the limitations of its accuracy as a consequence of the unique style of motion it has. An example of a delta frame can be seen in Figure 3.1. The last design considered was an A-Frame, which ultimately was picked for the 3D printer made at LMU. A-Frames provide a multitude of advantages over the other types of frames in the form of rigidity and durability. Rigidity improves the accuracy of machines by preventing the machine from shaking while it is printing. In terms of durability, the machine can be picked up and moved around without needing to be recalibrated due to the strength of the triangular designs on the sides of the printer. Figure 3.2 shows the A-Frame design.

Multiple extruder types were also considered for the design. A Bowden extruder takes the motor and extruder components and moves them from the X–Y carriage on to the frame. There is better control of the filament through reducing the weight of the X–Y carriage. This allows for fast, relatively accurate builds. However, the accuracy suffers in comparison to other

FIGURE 3.1
Delta frame.

FIGURE 3.2
A-Frame design.

products due to the springy nature of the filament running to the X–Y carriage, which does not allow for fine control of the amount of plastic being extruded. A multiple filament extruder was also a viable option. With a multiple filament extruder, there is not very fine control because both filaments are sent through the same nozzle and need to be controlled at the same time. In addition to this, though two different colors can be used in a build, between switching colors, there can be odd color mixes that may not satisfy the person producing the parts. The main advantage with a multiple filament extruder is the fact that there is only one head, significantly reducing the size of the X–Y frame.

Direct drive extruders were the third option considered. Direct drive extruders keep the motor and other extruder components on the X–Y carriage while driving the filament directly without any gearing. There is less fine control due to the excess weight on the carriage. This also requires a higher torque motor. However, by adding gearing to a direct drive extruder, there is no need for a higher torque motor. Though direct drive extruders may have some drawbacks with respect to accuracy, they are still very accurate in comparison to a Bowden extruder. Therefore, a geared extruder was picked for use within this design for accuracy. Geared extruders allow for finer control over the dispensation of plastic. Using this type of extruder in combination with a .25 mm nozzle, which is small for a 3-D printer, very accurate parts can be created at the cost of print speed. The gears used on the extruder also use herringbone teeth to prevent backlash from occurring, which would produce extra plastic when not necessary. A picture of a geared extruder can be seen in Figure 3.3.

Heated build plates are critical parts when it comes to printing certain materials such as ABS. A heated build plate prevents thermal gradients from forming within the printed material. Thermal gradients can cause deformation and warping, which is not ideal for the parts that need to be produced. A heated build plate can be seen in Figure 3.4.

FIGURE 3.3
Geared extruders.

FIGURE 3.4
Heated build plate.

This printer uses an Arduino mega 2650 combined with a ramps shield. The ramps shield contains 5 pololu stepper drivers to run the X-, Y-, and Z-axes in addition to two extruder motors. This will give future students the ability to expand and tune the new machine. An extra code can be added to increase the capability of the machine such as servo inputs, light inputs, and fan inputs. Many different features of the RP machine can be controlled via the Arduino.

3.4 Build Process

The 3D printer was first designed in Solidworks, a computer-aided design program, in order to show the compatibility of parts, size of the printer, logistics for construction, and to provide the builders with a reference. The Solidworks files operated as a blueprint for the construction of the printer and can be seen in Figure 3.5. The first step to construction was to cut down the 1.5 m V-slot stock on the horizontal band saw. There are 18 different V-slot extrusions in 6 different lengths needed for the whole assembly.

Using this model, the frame was constructed using V-slot aluminum, RP parts from the Makerbot Replicator 2, and varying fasteners. Extra fastening points were inserted into the frame for the various mounts for the Y- and Z-axes. The base was constructed first in order to provide a sturdy platform for the rest of the construction of the printer. The top half of the frame was then assembled and mounted on top of the base. The bottom bracket in Figure 3.6 holds the platform together and the top bracket in Figure 3.7 holds the top half of the frame to the platform.

There are four rapid prototyped mounting brackets that mount the A-frame to the top supports. These mounting brackets hold the top of the assembly together and can be seen in Figure 3.7. During construction, it was imperative to put T-slot nuts within the frame where more parts would be fastened, as T-slot nuts cannot be added or removed without disassembly. These T-slot nuts were used to fasten all of the brackets to the V-Slot. With the

FIGURE 3.5
CAD model.

FIGURE 3.6
Bottom bracket.

FIGURE 3.7
Top bracket.

frame constructed, the Z-axis motor supports a Y-belt drive motor, where it is mounted to the frame. The motor mounts and idler mounts are also designed and made with RP parts as seen in Figure 3.8a and b.

The next step was to build the frame for the Z carriage. This was assembled and mounted within the frame on the 2 mm pitch Z-axis screws. With the vertical supports somewhat loose in the frame the Z carriage can be used to move up and down to align the vertical supports. Once the vertical supports

(a)

(b)

FIGURE 3.8
(a, b) Motor mounts.

are aligned they can be tightened down to prevent them from moving. The Z-axis screws can be fastened to the Z-axis motors and mounted in the RP ends of the Z-axis carriage. The mounts for the V-slot wheels are slid into the supports and tightened down to stabilize the Z-axis carriage. The Y-axis carriage is composed of a ¼ in. acrylic with holes for V-slot wheels and acentric spacers to tension them. The acrylic also has holes to mount the spring suspended heated build plate. The final part of the axis assembly is the X-axis plate that has four V-slot wheels mounted in slots to give tension to them on the Z-axis carriage aluminum extrusions. Belts can be run for the X-axis and the Y-axis and zip tied to prevent them from slipping.

The extruders and extruder mounting plate were assembled next. The extruders work by gearing down the drive from the NEMA 17 stepper motor via herringbone gears to increase both torque and filament control. The gears drive hobbled bolts to create smallest diameter filament drivers. About 688 bearings that are spring loaded against the bolt act as idlers to force the filament against the hobbing on the bolts. The filament is forced through the hot ends using the stepper motors. Each hot end mounts into its corresponding extruder and has building instructions that can be found on the manufacturer website. The two extruders are mounted on the plate and the entire dual extruder assembly is mounted on the X carriage on four springs to allow for hot end leveling. The mounted dual extruder tips can be seen in Figure 3.9.

The next step is to mount the Arduino board, as seen in Figure 3.10, and start the process of wiring [3]. Each of the stepper motors has a corresponding location to plug. The heaters for the hot end and the relay input for the bed must both be properly secured to prevent short circuits and bad connections. The thermistors on each hot end and bed must also be plugged into their

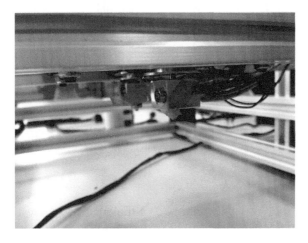

FIGURE 3.9
Hot-end extruders.

FIGURE 3.10
Arduino board.

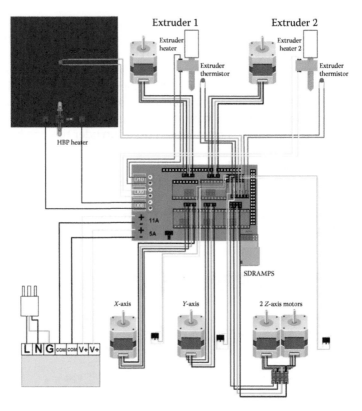

FIGURE 3.11
Wiring reference.

corresponding locations. The final part of the wiring was the 12 V rail of the advanced technology extended (ATX) power supply to provide power to the whole ramps board. All wiring can be referenced in Figure 3.11. Endstops need to be mounted adjustably to each frame axis. Endstops allow the printer to find the origin in 3D space. Special care must be taken when wiring the endstops to make sure they function properly. Endstops can be seen in Figure 3.12.

The 3D printer that has been designed and fabricated is shown in Figure 3.13.

3.5 Future Improvements

There are a multitude of different modifications that can be made to this design to further improve its ability to print. This solution to our design problem has not yet been evaluated, as the printer still needs to

FIGURE 3.12
Endstops.

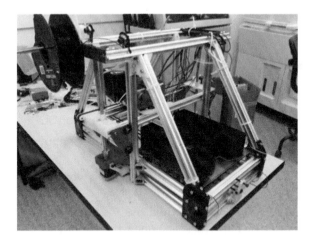

FIGURE 3.13
LMU 3D printer.

be calibrated to the appropriate settings. The printer is printing parts, but they are not as accurate as we would like them to be. The next improvement would be the development of an enclosed build chamber. By enclosing the print area, the parts printed within the heated volume of the printer will not warp easily, as they will be of uniform temperature throughout the printing process. This means that parts will only need to be printed once and prevent wasting material throughout the use of the printer. Another improvement that could be made to the machine would be increasing the size of the extruder nozzle. By increasing the size of the nozzle, faster dispensation of plastic can occur and create an overall faster build process for the 3D parts.

One major calibration task that must be done consistently with 3D printers is leveling. If the extruders are not level, the printer will print all parts on an angle and create useless parts. A major development in RP has been that of the automatically leveling extruder heads. This is a potentially large room for growth within the printer, as it would make the maintenance required to operate the printer significantly less. Some form of development with regard to the leveling of the machine would make it overall more accurate. Recently, patents have been filed within the field of RP with regard to subtractive processes. A difficult but rewarding task with regard to this 3D printer would be the development of a laser cutter for use in processes where a subtractive element may be necessary. This would greatly improve the flexibility of the design process allowing for more complex parts to be made. Last, the printer's esthetics need to be improved. As the printer sits today, there are a multitude of wires laying around it that need to be bundled and managed in such a way that they do not interfere with the printer's ability to print.

3.6 Questions

1. What is a RepRap project? What is its influence on 3DP?
2. What is a build platform? What are its main components? How do you improve adhesion of plastic to the bed?
3. What are the main components of 3DP? How would you select those components?
4. What kind of software you would use to control the precise operation of your 3DP? Can you select one from the open source? Is there any advantage of selecting an open-source software?
5. What are the different types of frame design you can consider for building your 3DP? Which one you would select and why?

References

1. T. Wohlers, Wohlers report 2016. 3D printing and additive manufacturing state of the industry. Annual Worldwide Progress Report.
2. C. Coward, *Idiot's Guide As Easy As It Gets 3D Printing*, Alpha by Penguin Group, New York, 2015.
3. K. K. Hausman and R. Horne, *3D Printing for Dummies*, John Wiley & Sons, Hoboken, NJ, 2014.
4. M. Frauenfelder, *Make: Ultimate Guide to 3D Printing*, Maker Media, Inc., North Sebastopol, CA, 2014, 1–112.

4

Calibrating the 3D Printer

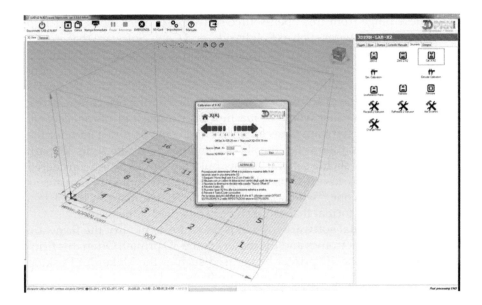

4.1 Introduction

The calibration of the 3D printer is vital for fast, easy, and quality prints. It is the mind of the 3D printer, telling the extruder where to go and what to do. In order to do this, several different softwares are used to program the 3D printer. Many of the softwares that are required for the calibration are open source, meaning that they are free. The field of open-source software is a growing community and has a strong following.

In this chapter, the many different types of software will be discussed. The pros and cons of specific software will be discussed. In addition, a step-by-step process of the changes in the software is also included. While this chapter covers only a specific software, other softwares can be used and instructions can be found online.

4.2 Types of 3D Printing Software

After building the 3D printer, the next step is to configure the software. There are several different types of softwares that are necessary for the 3D printer and the 3D component to be printed. The first type of software required is modeling software. 3D Modeling software will be used to create a part in 3D space. There are several different programs that can be used to create a part. These range from free open-source software such as Google Sketchup to commercial softwares such as SolidWorks and Catia. Software that creates 3D models is often referred to as computer-aided design (CAD) software. 3D models can also be downloaded online. A good place to download parts from is Thingiverse, a free online database of 3D designed models. After the part is created, another type of software is needed to slice up the object into layers. This is necessary because when the machine is printing, it prints one layer at a time. The slicing software takes each slice and converts it to a path for the hot extruder end to follow. All the slices combined form the solid 3D part. There is a variety of slicing software exists, but Cura and Slic3r are considered the most robust and easiest to use slicing software. The software transforms the part into a "g-code," which is a software language that can translate a sliced model into a format that can then be printed using the printer.

The third and final software needed is an editor to modify the firmware that controls the electronics and the motors of the 3D printer. Often this firmware comes with the 3D printer being used such as the MakerBot Desktop firmware, which comes with MakerBot products, and the PreForm firmware, which comes with FormLabs products [1]. However, there are some open-source firmware, such as Marlin, which can also be used. There are several motors that control the X-, Y-, and Z-axes, and another motor that controls the speed at which the filament is fed into the hot extruder head, all of which are controlled by the firmware. The firmware also tells the extruder head to heat up to a specific temperature. Firmware is able to read the g-code commands and transfer them to actual movements. It is necessary to be able to edit firmware to tailor it to a specific machine and tailor the settings.

4.3 3D Printer Software Configuration Using Marlin

In this section, Marlin, the software used for configuring a 3D printer to print a part, will be discussed. It is best suited for those who want to familiarize themselves with a general software that can be applied to multiple printers. Marlin includes various features such as arc support, dynamic temperature formatting, and SD card support among other features. An image of the Marlin logo can be seen in Figure 4.1.

FIGURE 4.1
Marlin software logo.

Once downloaded, Marlin codes must be altered to match the specific machine being used. This can be done manually, but often Marlin can pare with common 3D printer types (such as the MakerBot), and the code can automatically be altered.

4.3.1 Configuring MARLIN

To start configuring Marlin, select the configuration by tab. Under the tab, there is a long code. Changes to this code will make changes on how the 3D printer operates. The Marlin code is annotated very well, and it is extremely important to follow in order to make any necessary changes. The rate at which the machine communicates with Cura or Slic3r (the slicing programs) is the first variable that must be changed. It is called baudrate. A typical baudrate is 115200 when using Cura.

The next step is to tell Marlin which electronic board is being used to control the 3D printer (Figure 4.2).

FIGURE 4.2
Marlin code showing the baudrate and the electronic for a board.

The next step is to configure the thermistors and thermocouples on the 3D printer being used. It is important to follow the instructions in Marlin to input the correct information. Selecting the variable that works with the 3D printer is vital for proper prints to be made. The thermistors and thermocouples can be changed manually, as shown in Figure 4.3.

Following the temperature sensors, one must configure the endstops. Again, this is very important for the printer to know where the edge of a part is. This can be done automatically or coded manually, just as the temperature can be coded as seen in Figure 4.3.

When the motors are not being used they must also be disabled. The Z-axis is deactivated the most because it does not move while each layer is being completed. This is most often done manually through coding, where the movements are defined as "false." An example of this coding is "#define DISABLE_X false." If the direction of the axis needs to be reversed, it can be done through defining the inverse direction as "false" through manual coding [2].

After the endstop code is completed, the next step is to find out the maximum dimensions of the printer. It is important, so avoid reaching the limits of the axes in order to not burn out the motors or cause errors when printing outside of the print dimensions. The easiest way to know the limits of the

FIGURE 4.3
Marlin code showing the temperature changes.

machine is to do a homing test. From the home position, click on the button that says "move 10 mm" and count how many times it moves in each direction. If the machine is homing too fast, then lowering the values to compensate for the speed is important.

Finally, the most important variable is the axis steps per unit. This is very important to calibrate so that the machine moves exactly where it needs to go. If it is not calibrated correctly, then the extruder will be moving to places where errors could occur. There are four values in this portion of code in Marlin. The axis step per unit of the X-, Y-, and Z-axes as well as the steps per unit of the extruder. Each of the values for the axis steps per unit is calculated separately. For belt driven motors, the steps per unit is found by the number of steps per revolution, divided the spacing between the teeth on the gear controlling the motor and then divided by the pitch of the belt on the motor. The equation for this is shown below:

Steps per unit (belt motor) = (Motor steps per rev × belt pitch/gear teeth).

$$(4.1)$$

For threaded rods the calculation is different. The steps per unit are found by the motor steps per revolution divided by the pitch of the rod. Again, the equation for this is shown:

Steps per unit $(Z\text{-axis})$ = Motor steps per revolution/rod pitch. $\quad(4.2)$

For the geared extruder, the steps per revolution are found by the steps per revolution multiplied by the gear ratio (large to small), divided by the pinch wheel diameter times π. This is shown in Equation 4.3.

$$\text{Steps per unit} = \text{Motor steps per revolution} \times \frac{\text{Extruder gear ratio}}{(\text{Pinch wheel diameter} \times \pi)}.$$

$$(4.3)$$

These values can be calculated using an online solver at prusaprinter.org/calculator.

4.3.2 Testing the 3D Printer Movement

With the Marlin software downloaded and installed, it is time to test to make sure everything works. In order to do this, a series of programs called Printrun can be used. Printrun has several different programs, but in order to test the 3D printer, only the Pronterface program in Printrun is used. Pronterface is a graphical user interface that allows the user to jog the extruder head, heat the extruder, heat the bed, and test the motors. The layout of Pronterface can be seen in Figure 4.4. With Marlin installed and the 3D printed tested using Pronterface, the printer is ready to start printing parts.

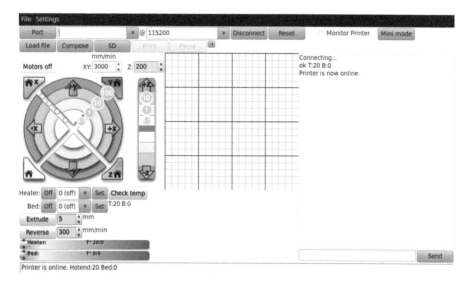

FIGURE 4.4
Screen shot of Pronterface software.

4.4 The First Print

This section will cover the printing process, starting with the design software. SolidWorks was used in this instance, but many other different types of CAD software can be used to print a part. The basic steps to print a part are downloading or creating a part in CAD, saving the part as an standard tessellation language (STL) file, importing the part to a slicing software, preparing the printer and homing the extruder (often done automatically), and finally printing the part.

These next sections will include steps for downloading parts, configuring Cura, and printing a part. The process is relatively the same for different types of software and different 3D printers.

4.4.1 Creating or Downloading a Part

Parts can be created on CAD software or downloaded from numerous websites. While it is easier to just download a part, learning CAD software will prove valuable for custom parts.

Thingiverse is a great place to download simple or complex parts alike if one does not want to create one themselves. There are plenty of other sites online from which parts can be downloaded. Another great source is 3DVIA or GrabCad. Once the part has been downloaded from one of these sites, or created in SolidWorks, it should be saved as an STL file. A picture of Thingiverse can be seen in Figure 4.5.

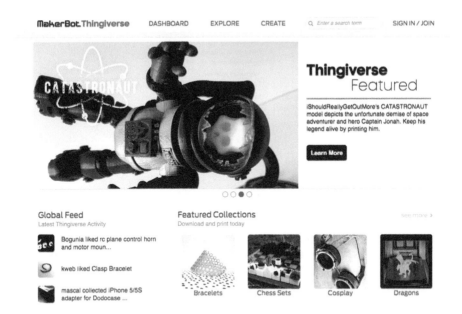

FIGURE 4.5
Thingiverse main website interface.

In order to slice the part, it sometimes needs to be imported to another program. Sometimes, certain printers, like the MakerBot Replicator, include slicing in the printer firmware. However, if the printer being used does not have this feature, another slicing program should be used. Two popular slicing programs are Cura and Slic3r [3]. Both Cura and Slic3r have their pros and cons. Cura has better user interface and is faster when it comes to slicing the part. It also allows the user to see the part and make changes in the program. The downside to Cura is that is does not always have the best quality and some prints are not as accurate as a result. Slic3r allows fine-tune adjustments to the print but the user interface is not as easy to use. Cura and Slic3r can be downloaded for free at the following sites.

4.4.2 Configuring the Cura Software

When Cura, a slicing interface from Stampante, is first run, it opens in the fast print mode. To take it out of the fast print mode, go to the normal mode tab and then select normal mode. To change the preferences to file, then select preferences. The preference window is shown in Figure 4.6.

There are several different settings that can be adjusted in Cura. These settings include the machine settings, filament settings, communication settings, and SD card settings.

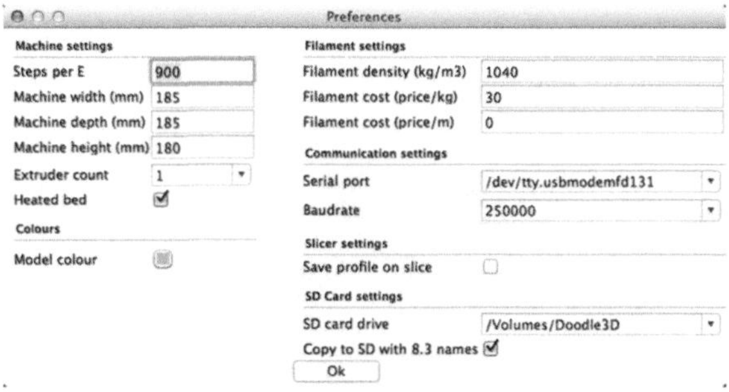

FIGURE 4.6
The Cura preference window.

Machine settings are the motor steps for the extruders, the dimensions of the 3D printer, the number of extruders, and a toggle box for whether or not there is a heated bed. The "Steps Per E" is the amount of steps per unit the extruder will go. This number will vary the amount of plastic that goes through the extruder.

Filament settings do not change anything on the 3D printer; instead, it is useful and provide information to calculate how much a part would cost when being printed. This value is based on how much plastic filament the part takes to be made. Through communication settings, Cura communicates to the printer. The baudrate, which was also set in Marlin, can be adjusted here. It also includes the serial port which will be connected to the 3D printer. SD Card settings lends the choice of the location of the SD to save files autonomously. It can also show short file names on the display of the printer.

4.4.3 Final Print Configuration

The final print configuration is the final step before printing a part. There are several inputs that can be adjusted that are detailed below [4]. The configuring interface is shown in Figure 4.7.

4.4.4 Accuracy

The first input that can be adjusted are the accuracy settings. This includes the layer height and wall thickness. Layer height controls the thinness of the layers of filament while the wall thickness controls the thinness of the outer walls of the part. The wall thickness should be two times the size of the extruder.

4.4.5 Fill

Fill is the second input that is in the final print configuration. It includes the bottom/top thickness and fill density. Bottom/Top Thickness: Controls the upper and lower bound thickness. These layers will have a complete infill.

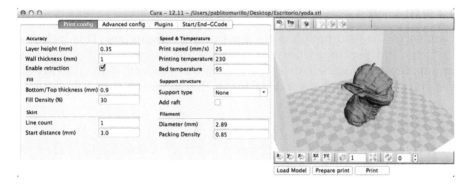

FIGURE 4.7
Cura final print configuration window.

Fill density is the inner density of the piece, similar to infill. If a part has 100% infill, it will be completely solid.

4.4.6 Skirt

Skirt is the next input that can be changed. Skirts encompass the part itself and allow for a more stable print. Underneath the skirt tab are the line count and start distance. Line count is the amount of lines printed around the part, which shows the outline of the part. This allows the user to know the size of the part. The start distance is the distance of the line around the part.

4.4.7 Speed and Temperature

Speed and temperature is the next section of the final print configuration. Print speed, print temperature, and bed temperature are the three sub-sections of this category. Print speed is determined by the speed of the motor as well as how accurate a print is desired. Printing temperature is the temperature that the plastic will be heated to in order to melt and extrude properly. Typically, the temperatures range from 185°C to 240°C depending on the plastic. Bed temperature is the temperature of the bed of the 3D printer. This is typically only necessary for acrylonitrile butadiene styrene (ABS) filament, which should have a bed temperature of 100°C to 110°C.

4.4.8 Support Structure

Support structure is another option that can be adjusted before printing. This includes the support type and raft. The support type will add support material to the part, which is encouraged if a part has large overhangs. Raft is a base layer for the part to be printed on. This is very useful for a part that will not stick to the base layer very well.

4.4.9 Filament

Filament is the final option of the final print configuration. This includes diameter and packing density. Diameter is the diameter of the filament in mm. Most filaments have a diameter of the filament on the packaging. Packing density is the correction factor for the amount of extruded plastic.

4.4.10 Printing

Once everything is set up, the "Prepare to Print" button can be clicked to generate the g-code. Once the g-code is generated, the "Print" button should be clicked. An alternative to this would be to save the g-code to an SD card. From the SD card, it can be plugged into the printer. Once in the printer, the file can be selected on the screen and printed.

Once the print has started, the process is over. If it is the first print, it is important to watch the 3D printer for a little while to make sure there are not any major mistakes. If the print is not as desired, then settings can be changed to make the print better.

4.5 Summary

This chapter covered the process on how to configure a 3D printer. To begin, software was chosen to program the 3D printer; in this case, Marlin was selected due to its many different functions. The Marlin code was then looked over, including the automatic and manual portions, which can be changed for the respective 3D printer. After Marlin is properly coded, Pronterface was used to test the movement of the 3D printer. With the 3D printer officially functioning, slicer software was then selected. Both Slic3r and Cura work, but in this case only Cura was covered. After making changes in Cura, a part can be downloaded or created. This part would then be sliced in Cura and imported to the printer. The printer can then print the part. If the part did not come out as desired, the options in Cura can be manipulated until a quality print is made.

4.6 Questions

1. What is calibration? Why is calibration so important for a 3D printer?
2. What is the difference between a proprietary software vs. open-source software?

3. What is Marlin? What are the special features of Marlin software? Would you use Marlin as a calibrating software and why?

4. What is Cura software? What is the difference between Cura and Marlin? Which software you would use and why.

5. What are the several parameters you need to control for calibration?

References

1. K.K. Hausman and R. Horne, *3D Printing for Dummies*, John Wiley & Sons, Hoboken, NJ, 2014.
2. Home, Marlin 3D printer firmware. Accessed September 24, 2016. http://www.marlinfw.org/.
3. J.F. Kelly, *3D Printing: Build Your Own 3D Printer and Print Your Own 3D Objects*, Que Publishing, Indianapolis, IN.
4. Stampante 3D Software Per Stampante 3D-Calibrazione Assi X In X2. 2016. 3DPRN. Accessed September 24, 2016. http://www.3dprn.com/software-per-stampante-3d-calibrazione-assi-x-in-x2/?lang=en.

5

Materials for 3D Printing

5.1 Introduction

Materials are the building blocks of 3D printers. The type of material directly affects the shape, dimensions, durability, and applications. In industry, all products are manufactured from one or multiple materials. For example, an automobile contains a wide variety of materials, such as steel for gears, glass for windows, plastic for dashboards, and rubber for tires. However, in 3D printing (3DP), there are not many materials available and the selection of materials is limited. The next generation of 3D printers should have improved processing methods so that a wider variety of materials with "intrinsic material" properties can be produced. Figure 5.1 shows a wheel prototyped for an automobile [1].

This chapter discusses the different types of materials currently being used in 3D printers. In general, there are three main categories of materials used in 3DP; they are liquid-based materials, solid-based materials, and powder-based materials. Each of these three categories has different types of materials such as polymers, metals, ceramics, and composites. These four types of materials will be covered in Section 5.2. The liquid-based, solid-based, and powder-based materials will be discussed in Sections 5.3, 5.4, and 5.5, respectively [2,3].

FIGURE 5.1
A car wheel is prototyped using additive manufacturing.

5.2 Types of Materials

This section describes the characteristics of polymers, metals, ceramics, and composites, which are the four main types of materials used in 3DP. More discussion will be devoted to polymers as more 3DP development has occurred in these materials. All of these materials will be discussed in terms of properties, advantages and disadvantages, and applications. Materials combination falls into any of the three categories, whether they may be metal polymers, ceramic polymers, or ceramic/metal composites. The three basic material types and their composites are shown in Figure 5.2.

5.2.1 Polymers

Polymers are the most common type of materials used in 3DP. They contain structural units called *mers*, which form to create the polymer. Some general properties of polymers include low electrical and thermal conductivity and high strength-to-weight ratio, which make them particularly useful in 3DP. Polymers can also be processed at low temperature due to their low glass transition temperatures. Also, they have low density and good chemical corrosion resistance. Polymers can be divided into three different categories: thermoplastic polymers, thermosetting polymers, and elastomers.

5.2.1.1 Thermoplastic Polymers

Thermoplastic polymers are the most popular type of materials used for 3DP. They make up most of the types of filaments used in fused deposition

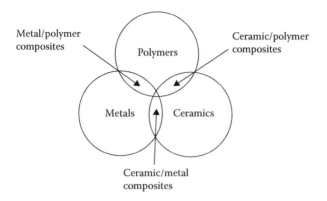

FIGURE 5.2
Basic material types with their composites.

modeling with variety of colors. Thermoplastics are unique because they can withstand multiple heating and cooling cycles without altering their molecular structure. Common thermoplastics include *polylactic acid* (PLA) (Figure 5.3), acrylonitrile butadiene styrene (ABS), polycarbonate (PC), polyamides (nylon), polyvinyl alcohol (PVA), polystyrene (PS), and polyethylene (PE). Table 5.1 summarizes the advantages and disadvantages of different thermoplastic types.

5.2.1.2 Thermosetting Polymers

Thermosetting polymers are arranged in a cross-linked 3D shape. They are chemically transformed into a rigid structure after cooling from a heated plastic condition. However, a thermosetting polymer cannot be

FIGURE 5.3
PLA thermoplastic used as a filament for a 3D printer.

TABLE 5.1

The Advantages and Disadvantages of Thermoplastics

Thermoplastic	Advantages	Disadvantages
PLA	Variety of colors Environmental friendly (biodegradable) Smooth appearance and finish No harmful fumes during printing High print speed and resolution	Brittle Less flexible Low heat resistance Slow cooling down
ABS	High strength Flexible High heat resistance High melting point Easy to extrude	Notable smell Not suitable for foods (nonbiodegradable) Production of airborne microparticles
PC	High resistance to scratches and impact High strength High durability High heat applications	High extruder temperature required State change when exposed to UV light
Polyamides (nylon)	Excellent layer adhesion Variety of colors Very strong Self-bonding and lubrication	Sensitive to moisture Nonbiodegradable Need to dry before printing
PVA	Recyclable Conductive Excellent film formation	Expensive Water soluble
PS	Low-thermal conductivity Cheap Recyclable	Poor strength Nonbiodegradable
PE	High resistance to impact Recyclable	Difficult to bond Poor strength

reshaped once it is cooled during the curing process; therefore, they cannot be recycled. In addition, in some cases, the polymers used are cured by nonthermal mechanisms. This type of polymer typically includes alkyds, phenolics, amino resins, and epoxies. Thermosetting polymers are brittle, possessing mechanically and thermally better properties than thermoplastics. That is, they are stronger than thermoplastic polymers, making them suitable for high-temperature applications. Thermosetting polymers have a wide range of applications, such as electronic components made from alkyds, due to their good electrical insulation. Also, phenolics have high heat and electrical resistance, making them great electrical wiring and connector devices. Aminos are known for their rigidity and hardness; therefore, they are used in housing appliances such as toilet seats. Lastly, epoxies have great dimensional stability and possess high-mechanical strength properties; they are typically used in pressure vessels, rockets motor casings, heavy tools. Figure 5.4 shows a comb that has been 3D printed and ready to use.

FIGURE 5.4
A fully functional 3D printed comb.

5.2.1.3 Elastomers

Elastomers are based on polymers in which they exhibit significantly elastic behaviors, hence the name. Elastomers are mainly made from a chain of carbon, hydrogen, and oxygen atoms connected in a cross-linked structure. Elastomers are famous among polymers in their wide applications and include natural rubber, neoprene, silicone, and polyurethane. The unique mechanical properties that elastomers possess, such as elasticity and resilience, make them versatile in many applications. Essentially, the elasticity and resilience, which is defined as the ability of a material to return to its original shape after being deformed or stretched by an applied load. In addition, elastomers possess good thermal and electrical insulation that are demanded in many engineering applications. In 3DP, elastomers are used to grow gaskets, seals, and hoses.

5.2.2 Metals

Metals are another type of materials used for 3DP. Metal alloys are composed of two or more elements, in which at least one is a metallic element. Alloying elements are added to the metal alloy to improve their physical properties. There are two types of metals: ferrous and nonferrous.

Ferrous metals are based on iron element. There are thousands of varying types of ferrous metals, but only few of them are used in 3DP. The most common ferrous metal is steel and cast iron. The essential ferrous alloying element is carbon from which steel and cast iron are formed. The carbon quantities added to the alloy varies from an alloy to another. The range of carbons added is usually between 0.1% and 0.7% of the compound. Some other common alloying elements are chromium, manganese, nickel, and molybdenum. Each alloying elements enhances particular property of the alloy. Some general properties of ferrous alloys include good strength, relatively low cost among metals, and process ability.

Nonferrous alloys include all other metallic alloys. Nonferrous metals include aluminum, nickel, copper, gold, magnesium, silver, tin, zinc, and titanium. Some nonferrous alloys are easy to process and others not. For example, aluminum alloys are easy to process, but titanium and nickel alloys are not. This is why aluminum alloys are widely used in the aerospace industry. In rocket engine and gas turbine, for instance, nonferrous alloys with high-temperature performances, such as nickel and cobalt as the metal base constituent alloy, are needed due to their lower tendency for oxidation. Figure 5.5 shows a car engine made from steel.

Metals have various properties which make them attractive in 3DP industry. They provide a good combination of strength, toughness, and machinability unlike ceramics, which are brittle, and polymers, which cannot be used in high-temperature applications. In addition, metals have good electrical and thermal conductivity and most importantly good ductility.

5.2.3 Ceramics

Ceramics are compound that contain metallic (or semi-metallic) and non-metallic elements. Some common metallic element of ceramics are Al, Mg, Si, and Zr. Typical nonmetallic elements are oxygen, nitrogen, carbon, and boron. Some characteristics of ceramics include low electrical and thermal conductivity, high strength, high melting point, and high brittleness. An example of that would be diamond, an allotropic form of carbon. Diamond is considered to be a ceramic because of its high hardness and brittleness. In addition, inorganic glasses such as SiO_2-based compounds are often considered to be ceramics due to similarity in physical and chemical properties.

Modern ceramics include some materials such as alumina, carbides, and nitrides. These materials enhance the properties of ceramics and allow them

FIGURE 5.5
Honda GHX 50 engine is made from steel.

to be useful as cutting tools and grinding abrasives. Ceramics can be divided into two groups: crystalline ceramics and glasses. The crystalline ceramics are usually formed from powders and then sintered, while glass can be melted and later formed in processes such as traditional glass blowing.

5.2.4 Composites

Composites are modified nonhomogeneous materials. They could be mixtures of polymers, ceramics, and metals. A composite is a material that consists of at least two phases that are bonded together to yield in superior properties compared to the original properties. The term composite refers to a matrix material, in which fibers or particulates, such as Al_2O_3, are embedded. Embedding high modulus fibers or particulates into a lower modulus matrix improves the properties of the host matrix. The matrix provides the composite material with greater elasticity and strain-to-failure, whereas the fibers provide enhanced strength. The reinforcing phase improves the fracture toughness of the composite matrix, whereas the fibers act as an extra barrier during extension and fracture.

Most fibrous composites are constructed in thin layers with long fibers laid down in a singular direction, alternating that direction in successive layers to ensure that the composite material is strong in all directions by using laminate object manufacturing (LOM) process. Some composites resemble plywood with strength in only one direction, while others weave the fibers into a 3D structure. The fibers can be either short fibers (i.e., particles or flakes) made of glass, carbon graphite, aluminum oxide, or silicon carbide, or long fibers such as the tungsten-boron filaments. The matrix material is made from a ceramic such as silicon nitride, or high-temperature plastic such as an epoxy resin, or from a metal such as aluminum. The most common type of composite in use today is fiberglass which is glass fiber-reinforced polymers due to its durability and light weight. Table 5.2 shows some advantages and disadvantages of composites.

In fact, it is possible to achieve a combination of properties not possible for any one of the original three types of materials (i.e., plastic, ceramic, or metal). An example of a famous composite that is widely used in industry and heavy duty applications is carbon fiber, it is known for its light weight and has excellent strength-to-weight ratio. Figure 5.6 shows carbon fiber used in manufacturing tubes.

TABLE 5.2

Advantages and Disadvantages of Composites

Advantages	Disadvantages
High strength-to-weight and stiffness-to-weight	Expensive to make
Greater toughness and fatigue properties than metals	

FIGURE 5.6
Carbon fiber tubes.

5.3 Liquid-Based Materials

5.3.1 Polymers

Liquid-based materials are thermoplastic polymers (polyamide nylons) or thermosetting polymers such as epoxies. In stereolithography, parts are built from a photosensitive polymer fluid that cures under exposure to a laser beam. This process has seen much development in the types of materials used, especially in photosensitive polymers or photopolymers. Stereolithography has generally been used on acrylates and epoxies. However, it is expected that stereolithography resin suppliers will continue to make progress in creating new materials that have selected thermoplastic properties.

In the late 1960s, the first UV curable photopolymers were developed to reduce air pollution from solvent-based coatings. Photopolymers are solidified (cured) when exposed to electromagnetic radiation with a specific wavelength including gamma rays, X-rays, ultraviolet (UV), visible light and infrared. Radiation technology today uses electron-beam (EB), and UV curing of photopolymers as the most common commercial applications.

Acrylate-based photopolymers are the most widely used resin systems developed for stereolithography. Later, resins were developed based on vinyl-ether (e.g. Allied Signal's Exactomer resins) and epoxy systems. Most resins in use today are epoxies. These resin systems are developed to react to UV light in the 325 nm laser wavelength from a helium-cadmium (HeCd) laser for the SLA-250 stereolithography system [3D Systems]. Other resin systems have been developed for different laser wavelengths. DSM Desotech has been in existence as an innovative manufacturer of UV/EB-curable materials, and the new division is called DSM Somos.

Perfecting the characteristics of resins has taken a long time. The early resins had very high shrink factors and low impact strengths after curing. The larger the shrink factor, the more difficult it becomes to create an accurate model. For instance, an early acrylate resin, released by Ciba-Geigy in 1988, had a shrink factor of 0.8%–1.1% and a low-impact strength. The next major breakthrough in resin development was SL5143, released in 1991. SL5143 was also an acrylate resin, with a shrink factor of 0.7% and an order of magnitude higher impact strength. The increase in impact strength was vital because early models would shatter if they were accidentally dropped. SL5143 was the first resin with enough strength to allow normal handling without the fear of breaking the model. The reduction in shrink factors was also very significant, and the most recent resins have 0.4% or less.

5.3.2 Metals and Composites

Because liquid-based materials are mainly thermoplastic polymers, currently, there is not that much variety to other materials such as metals. However, composites such as carbon fiber can be melted into a liquid form and inject molded similarly as plastic using a new technology that is still new to the 3DP world.

5.4 Solid-Based Materials

Solid-based rapid prototyping (RP) systems use solids as the primary medium to form prototyped parts. These RP systems include Stratasys' fused deposition modeling (FDM) using materials such as PC, ABS, polymethyl methacrylate thermoplastic, wax, and elastomer. Also, other RP system that utilize solid materials are Cubic Technology's laminate object manufacturing (LOM) (paper coated with a proprietary heat-activated adhesive) and Solidica's ultrasonic object consolidation (UOC) [4].

5.4.1 Polymers

Stratasys' FDM offers a unique variety of thermoplastic modeling materials. All these materials are eco-friendly and altered using existing thermoplastics and waxes. The FDM process uses a spool-based filament system to feed the material into the machine while heat is added. The material options currently include ABS materials including *ABS (P400)*, a high-impact grade of ABS called *ABSi (P500)*, investment casting wax (ICWO6), and an elastomer (E20). These are all thermoplastics that soften and liquefy in while heat is added, then are deposited in 2D layers. In 2002, Stratasys developed two other new thermoplastic materials—PC and polyphenylsulfone (also called polyphenylene sulfide or PPS). These two new materials uniquely possess heat-deflection temperatures of 125°C and 207°C, respectively yielding in higher results from original thermoplastics.

The use of ABS is essential in applications that require impact resistance, toughness, heat stability, chemical resistance, and the ability to perform functional tests on sample parts. ABS is considered a carbon-chain copolymer that can be made via dissolving butadiene-styrene copolymer in a mixture of acrylonitrile and styrene monomers, and then polymerizing the monomers with free-radical initiators. Three structural units of ABS, that is (A, B, and S) have a balanced set of properties, the acrylonitrile providing provides heat resistance, on the other hand, the butadiene groups impart good impact strength, and the styrene units give the copolymer its rigidity property.

Stratasys' other ABS material mentioned above is *ABSi (P500)*. ABSi (P500) is a special medical-grade of ABS that meets all FDA Class VI requirements, i.e., it can be sterilized with gamma radiation, and is resistant to medical environment that is high in chemicals contact involvement. Another FDM material is *Elastomer (E20)*. This thermoplastic polyester-based elastomer has been developed for applications where flexible material must possess toughness, and durability in such applications as seals, gaskets, bushings, hoses and tubing. In order to enhance mechanical and thermal properties of Elastomer (E20), such as the strength, toughness, and temperature capability of solid-based RP thermoplastics, PC and polyphenylsulfone (PPSF) were developed. In fact, PPSF is resistant to contacting gasoline, anti-freeze, and sulfuric acid. The new Genisys FDM machine created a thermoplastic polymer known as polyester (P1500). Polyester is known for having good mechanical, electrical, and chemical properties as well as good abrasion resistance. Common applications include gears, cams, pumps, and rollers.

5.4.2 Metals

Solidica uses a process that is called UOC, this process uses ultrasonic energy in order to create a solid state bond between metal sheets. The ultrasonic energy is imparted to the interfaces of the metal sheet, causing frictional heating across a ~20 μm interface, the atoms are able to diffuse across the interface to provide a strong metallurgical bond. The process is ideally suited for thin foils ~25 μm (~0.001 in.) of material. However, thicker sheets can also be used. The major advantage of this process over most RP processes is the lack of liquid to solid phase change in the materials involved.

5.4.3 Composites

Composites are basically mixtures of dissimilar materials. Composites are structured from polymer–matrix, metal–matrix and, ceramic–matrix composites. The matrix of a composite defined as is the continuous, surrounding phase where the fibers or particulates are imbedded. The matrix usually has a lower modulus than the fibers; therefore, this property allows the transfer of the load to the high-strength fibers. Composites are popular in the aerospace industry, because they have high strength and are light weight, in other words, the term is strength-to-weight for scientific use. The method of structuring composites involves manual lay-up of 2D fiber laminates that consume a lot of time. Cubic Technologies (formerly Helisys) was the first organization to automate the lay-up process using LOM in the RP industry. Here are the two composite material examples for the use of 3DP beginning with LOM that uses the method of lay-up, consisting of organic fibers in a 2D laminates of paper shape. These organic fibers are coated with a proprietary heat-activated adhesive (epoxy binder). The laminate is then dispensed from a roll that is 114 µm (0.0045 in.) thick and is bonded by a curing process of the the epoxy binder. Another example of a composite material is 3D Systems (formerly DTM Corporation) which is a glass-filled (GF)/nylon composite that is called Duraform GF.

5.5 Powder-Based Materials

Powder-based RP systems use powder as the prime material to prototype parts. In this section, powder-based materials will be discussed in relation to the following RP systems: selective laser sintering (SLS) of polymers and composites, direct metal deposition, direct shell casting for polymers, metals, and ceramics.

5.5.1 Polymers

5.5.1.1 Thermoplastics

SLS process utilizes the Sinterstation (or the newer version called Vanguard) from 3D Systems, which is used to fabricate both polyamide (nylon), which is called DuraForm PA and PC. The laser-sintered method raises the density of these materials to an intermediate density and then postprocessed to increase their density.

5.5.1.2 Polymer Composites

Duraform GF is a glass-filled/nylon composite that is composed of a glass-filled SLS polymer–matrix composite. Duraform consists of glass particles imbedded inside an SLS nylon matrix, which work on lowering the tensile strength of SLS nylon. In addition, the glass particles make the matrix more brittle than the SLS nylon. However, this composite has a higher elastic modulus, which adds a closer match to the modulus of an actual component during the test inside a wind tunnel.

5.5.1.3 Elastomers

DSM Somos 201 is a powder, a thermoplastic elastomer that is sintered to create highly flexible parts with elastomer characteristics. These models can be used in place of urethane, silicone, or rubber parts in such applications as moldings, gaskets, hoses, and vibration dampers. Somos 201 has a melting point of 156°C that makes it withstand heat and chemical solvents.

5.5.1.4 Powders

The process of SLS involves heating powders slightly below the melting point using a laser in the case of a metal. Here, the temperature is high enough to sinter or bond the individual powder particles together. Thereafter, the post-processing is used to increase the density of the powder.

The size of the powder particles in SLS is very small. For instance, the large-sized particles cause the surface of the part to be very course. If the particles are too fine, then the surfaces of the particles develop electrostatic charges that make the powders difficult to spread in a 2D layer. In addition, larger particles need a faster sintering rate.

5.5.1.5 Selected Properties

SLS materials are shown with their properties in Table 5.3. It should be noted that the mechanical properties of the SLS polymers (polyamide and polycarbonate) are much less than those for the FDM, PC and PPSF.

5.5.2 Metals

The main focus in RP materials is in the area of powder metallurgy where significant advances have occurred in the last few years.

5.5.2.1 Selective Laser Sintering

When Dealing with SLS process in the Sinterstation (or Vanguard) equipment, it is important to understand that the CO_2 laser does not sinter the metal particles together. Moreover, the laser sintering is not designed to heat the metal powders to ~$0.5T_{mp}$ (about half their melting point), which is required for the sintering of metal powders. The thermoplastic binder is used

TABLE 5.3

Properties for Selective Laser Sintering of Powder Materials

Property	SLS Materials				
	DuraForm PA Polyamide	DuraForm GF Glass/ Polyamide	Polycarbonate	LaserForm ST-100	Somos 201 Elastomer
Tensile Strength (MPa)	44	38.1	23	510	N/A
Elastic Modulus (GPa)	1.60	5.91	1.22	137	20
Elongation at Yield (%)	9%	2%	5%	10%	111%
Impact Strength (notched Izod), (J/m)	214	96	53	N/A	N/A
Hardness (Shore D-scale)	N/A	N/A	N/A	87 (Rockwell B)	81 (Shore A-scale)

to coat the metal particles while the binder of powders is sintered together to form fragile green parts. The postprocessing usually involves burning-out the polymer binder and infiltrating the metal powders with a lower melting liquid metal which is usually bronze or copper. SLS metal powders used are: steel, 420 stainless steel (ST-100), and 316 stainless steel. For example, the LaserForm ST-100 binder-coated metal powders have a particle size of ~36 to 76 μm. After laser sintering and binder burn-off, the stainless steel powder is 40% dense (or 60% porous). Afterwards, a liquid infiltrating is added to postprocess the steel powder with commercial bronze. Finally, the material possesses similar properties to P20 tool steel, where it is used for tooling in injection molding machines.

5.5.2.2 SLS and Hot-Isostatic Pressing

Newer processes have been developed for densifying steel powders rather than using the method of liquid metal infiltration. For example, hot iso-static pressing (HIP) compacts SLS powders to full density using high pressures (~100 MPa) and temperatures (~1100°C). In SLS/HIP process, the laser beam fuses the metal powder only on surfaces to form gas impermeable skin around the part that exceeds 92% theoretical density. Afterwards, the part is then evacuated inside a chamber and postprocessed to near net-shape by "container-less HIP" to full density. The advantage of this process over conventional HIP is that the part does not have to be placed in a container which eliminates the adverse part-container reactions. Materials that use this method are steels, cermets, titanium and its alloys, and refractory metals.

5.5.2.3 Direct Metal Laser Sintering

Consolidating metal powders to nearly full density has been developed by Electro-Optical Systems in Germany where no liquid metal infiltration is required. The process only uses a laser-sintering process, but the DirectSteel 20-V1 powder achieves approximately 95% density. The powders used are 20 µm from materials such as steel, bronze, and nickel.

5.5.2.4 Direct Metal Deposition

There are three organizations that have developed under the direct metal deposition process: Optomec, Precision Optical Manufacturing (POM), and AeroMet. Direct metal deposition is a process that injects metal powders into a melted pool on a substrate surface, while laser scans the geometry layer by layer. While Optomec uses a process that is called laser engineered net shaping, both POM and Aeromet call their process laser additive manufacturing. These processes look very similar to a net shape, and usually require a final machining to obtain a fine finish. The surface finish averages ~200–500 µin.

All the processes involve the creation of metal parts using either a Neodymium: Yttrium Aluminum Garnet laser for Optomec, or a CO_2 laser for POM and Aeromet.

5.5.3 Ceramics

Soligen's ceramic materials make direct investment casting shells (or molds) without the use of wax patterns. Their method called Direct Shell Production Castings (DSPC), builds the ceramic shell layer by layer. Afterwards, sintering the layers of the ceramic shell and molten metal is poured into the shell to finalize the part. For instance, engine cylinder heads of aluminum, steel alloys are some materials that Soligen's ceramic shells have casted.

5.5.3.1 Aluminum Oxide

In the DSPC process, a fine layer of alumina (Al_2O_3) powder is distributed using a roller mechanism on a new building platform. Using MIT's 3DP process, a liquid binder of colloidal silica (SiO_2) is grown onto a bed of alumina powder. After sintering, the silica bonds the alumina particles forming a rigid structure. DSPC process can create a ceramic part directly rather than using it as a shell for a casting mold.

5.5.3.2 Zirconium Oxide

Zirconia is being developed at Soligen, this material provides an increased cooling rate, similar to sand casting. Zirconia possesses resistance to thermal shock, wear, and corrosion in addition to a low-thermal conductivity property. Zirconia (ZrO_2) transforms from a tetragonal to a monoclinic structure when it cools down from an elevated temperature. This transformation may initiate or propagate cracks in the part yielding in a possible failure of the material. However, adding CaO, MgO, or Y_2O_3 to zirconia will stabilize the cubic phase throughout various temperatures, thus avoiding the destructive phase transformation and creating a partially stabilized zirconia (PSZ). This PSZ ceramic also has enhanced strength, toughness, and reliability than the unstabilized zirconia. A newer development for Zirconia is called transformation-toughened zirconia (TTZ), which improves on PSZ by providing enhanced toughness. Table 5.4 below shows the general ceramic properties for alumina, PSZ, and TTZ.

5.6 Common Materials Used in 3D Printers

5.6.1 PLA

PLA is one of the most common thermoplastic materials used in 3DP because it is environmentally friendly and is a biodegradable polymer created from sugar plants such as tapioca, corn, and sugarcane. It also does not require a heated build plate because PLA melts at a very low temperature. Its melting point is at around 160°C, but bonds better at 180°. PLA is also much more brittle than other thermoplastics, however, some improvements of it are being made in order to increase its flexibility and reduce carbon footprint during the creation process. One of the reasons why it is so popular is because of its ease of creation from whatever natural sugar plants which are locally available [5].

5.6.2 ABS

ABS plastic is used in a variety of ways from industrial applications for extrusion to children's toys such as Lego bricks. Therefore, its properties are well known and the quality of filament is easily controlled during manufacturing.

TABLE 5.4

General Ceramic Properties for Alumina, PSZ, and TTZ

Property	Alumina	PSZ	TTZ
Tensile Strength (MPa)	210	455	350
Flexural Strength (MPa)	560	700	805
Compressive Strength (MPa)	2100	1890	1750
Elastic Modulus (GPa)	392	210	203
Fracture Toughness (MPa $\sqrt{m}$)	5.6	11.2	12.3
Density (mg/m³)	3.98	5.8	5.8

ABS is known for its higher melting point than PLA, easy extrusion with less friction as it passes through the extruder, able to be printed on Kapton tape or a thin layer of ABS cement. Because ABS shrinks as it cools, a heated build plate is necessary in order to prevent earlier layers from contracting and avoid warping large objects. It should also be noted that if ABS is used in confined spaces, a mild odor during extrusion can affect chemically sensitive birds and people. Without adequate filtration, ABS can produce more airborne microscopic particles.

5.6.3 PC

Polycarbonate materials are a recent addition to the filaments of 3DP. They have a variety of uses in CDs, DVDs, and automotive and aerospace industries due to its high resistance to scratches and impacts. PC plastics have also been used in the creation of "bulletproof glass," however, when used in 3DP, the layering creates microscopic voids between the layers, so the final result is not stronger than molded industrial equivalents. Even though PC material is known for its high strength and durability, the required temperature for its extrusion is 260° or higher, which some 3D printers cannot sustain. These types of objects can also undergo a change in state when it is exposed to ultraviolet light, by becoming more opaque and brittle.

5.6.4 Polymides (Nylon)

Nylon is another recent addition to 3DP that has become useful due to its flexibility and strong self-bonding between layers, which allows it to have excellent layer adhesion. The temperature required for extrusion is between 240°C and 270°C. Nylon is also resistant to acetone, which dissolves ABS and PLA. Due to its excellent layer-bonding aids, nylon is used to produce good flexible vessels such as vases and cups.

5.7 Materials Selection Considerations

There are a couple of considerations that need to be taken into account when looking for the best material that fits your need. These considerations include application, function, geometry, and postprocessing. The following are brief descriptions of each:

5.7.1 Application

When choosing a material for a certain project, it is important to ensure that the material can deliver the required certifications for the project.

For example, some projects require 3DP materials that offer sterilization capabilities, biocompatibility, FDA certifications, fire-retardant certifications, chemical resistance, or other certifications that may be critical.

5.7.2 Function

3DP materials are subjected to extreme testing in order to answer the kinds of stresses it can endure and the level of taxing environment the material will excel in. The ability of a material to function in a desired application relies partly on design.

5.7.3 Geometry

It is important to consider the dimensional tolerances, design wall thicknesses, and minimum feature execution when choosing a material for 3DP. 3DP materials are often inseparable from their corresponding technology, which each has unique geometric executions, whether it is FDM, Sterolithography, or SLS.

5.7.4 Postprocessing

Some 3DP materials may be better suited to certain postprocessing methods than others. For example, heat treating stainless steel versus postcuring a photopolymer [6].

5.8 Summary

The 3DP materials (polymers, metals, ceramics, and composites) were reviewed in terms of characteristics, pros and cons, and applications. Polymers are the most common type of materials used in 3DP due to their low electrical and thermal conductivity (insulators), low density, and high strength-to-weight ratio. Polymers are classified into three categories: thermoplastics, thermosets, and elastomers. Examples of thermoplastics include PE and PC. Epoxies and phenolics are examples of thermosets while rubber is an example of elastomers. Metals, on the second hand, have a good combination of strength, toughness (ductility), thermal and electrical conductivity, and machinability, unlike polymers which cannot be used in high-temperature applications. The most common 3DP metals are steels, titanium alloys, aluminum alloys, and nickel-based superalloys. These metals are mainly fabricated by an SLS, direct metal deposition, and UOC. Ceramics, on the other hand, have high strength, which make them useful for cutting tools. The most common 3DP ceramics are Al_2O_3, SiO_2, and ZrO_2. Finally, composites are mixtures of polymers, metals, and ceramics, which make them have various properties. They are mainly classified into polymer–matrix, metal–matrix and ceramic–matrix composites. 3DP composites have been fabricated SLS, UOC, and LOM.

5.9 Questions

1. What is the difference between parts made from thermoplastic or thermoset? Please explain all of your ideas.
2. What are the major differences between the mechanical properties of plastics and metals?
3. Mention several advantages and disadvantages of Polylactic acid (PLA), Polymides (nylon), Polystyrene (PS), and Polyethylene (PE).
4. What is composite? What are the advantages and disadvantages of composites?
5. Mention several liquid-based materials. Which one is most frequently used?
6. What is the most popular material for powder-based 3DP? Mention some differences between DuraForm PA and Somos 201 Elastomer.
7. What mechanical properties do elastomers have that thermoplastics do not have?
8. Discuss several parameters you would consider for material selection for your printer.

References

1. http://www.thingiverse.com/thing:39751. Courtesy of Barspin, RC Truggy Rims. December 30, 2012.
2. S. Kalpakjian and S.R. Schmid, *Manufacturing Engineering and Technology*, Prentice Hall, Upper Saddle River, NJ, 2001.
3. D.R. Askeland, *The Science and Engineering of Materials*, PWS Publishing Company, Boston, MA, 1994.
4. R. Noorani, *Rapid Prototyping Principles and Applications*, Chapter 6, John Wiley & Sons, Hoboken, NJ, 2006, 156–195.
5. K.K. Hausman and R. Horne, *3D Printing for Dummies*, John Wiley & Sons, Hoboken, NJ, 2014
6. Stratasys, 3D printing & advanced manufacturing. 3D printing materials: Choosing the right material for your application. https://www.stratasysdirect.com/content/white_papers/STR_7463_15_SDM_WP_3D_MATERIALS.PDF.

6

Classifications of Rapid Prototyping and 3D Printing Systems

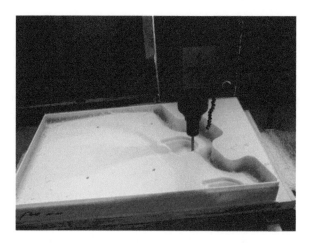

6.1 Introduction

The rapid prototyping (RP) and 3D printing (3DP) systems can be classified in a variety of ways depending on the physics of the process, the source of energy, type of material, size of prototypes, etc. The most widely accepted means of categorizing RP and 3DP systems in the industry is by determination of the initial form of the raw material. This text will follow the accepted norm. Accordingly, all RP and 3DP systems can be categorized into solid-based, liquid-based, and powder-based systems. For the purposes of this book, a classification scheme based on the initial form of the material is presented. Accordingly, all RP and 3DP systems will be classified as solid-based, liquid-based, or powder-based systems [1].

In this chapter, we shall study the solid-based system that includes the fused deposition modeling (FDM) process, liquid-based systems that include the stereolithography (SLA) process, and powder-based system that includes the selective laser sintering (SLS) processes. Most other companies use one of these fundamental processes for their printers. For each system, information on company profile, the principle, the process, the products, advantages and disadvantages, and application will be provided.

6.2 FDM Systems

6.2.1 Stratasys RP Machines

In 1988, Scott Crump, the President and CEO of Stratasys Inc., developed the FDM process. Stratasys's mission was to provide design engineers with cost-effective, environmentally safe, rapid modeling and prototyping solutions. FDM generates 3D models using an extrusion process.

Using patented FDM and PolyJet RP processes, Stratasys RP systems create precision 3D prototyping parts directly from the 3D CAD systems for use in testing form, fit, and function throughout the design and development process.

6.2.2 Principles of FDM

Rather than being tooled from a solid body, the 3D models made by RP technology are built by the addition of materials and/or the phase transition of materials from a fluid to a liquid state.

While many different approaches exist in the methodology of RP, this section focuses on FDM. In this method, a model created in a CAD program is imported into a software program specifically designed to work with the FDM machine, in this case, Insight. The data for the model are saved as an STL (stereolithography) file for interpretation by Insight [2].

This program then uses a slicing algorithm to divide (or slice) the computer graphical solid model into layers (Figure 6.1), each representing a layer of the

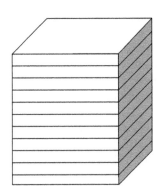

FIGURE 6.1
Sliced model.

material to be used. The information for the model is then downloaded to the FDM machine as a Stratasys Machine Language (SML) file. The machine builds the prototype by extruding a thin stream of semi-liquid acrylonitrile butadiene styrene (ABS) plastic, referred to as a road, from a heated head or tip. The road is laid across the X-Y plane one layer at a time, each layer corresponding to a slice determined by the Insight program. Each slice has the same height as a road.

To build each layer, the extrusion head deposits the outline of the layer first, and then fills each layer according to the raster dimensions set by the operator. The raster fill of the outline is a series of straight line segments that link together and are uniformly spaced (Figures 6.1 and 6.2). As each layer is deposited, it fuses with the previous one, creating a solid model.

The ABS P400 plastic used is a durable acrylonitrile-butadiene-styrene-based material that is appropriate for concept models as well as for testing of form, fit, and some function. The ABS P400 plastic is impact resistant, has a relatively high tensile strength, and is heat, scratch, and chemical resistant. This material has a high thermal expansion for a plastic and is lower in cost than most engineering thermoplastics. However, ABS plastic has limited weather resistance and is not very resistant to solvents [3]. Table 6.1 shows some of the advantages and disadvantages of ABS plastic.

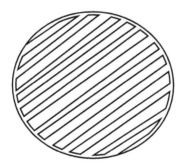

FIGURE 6.2
Contour and raster.

TABLE 6.1

Advantages and Disadvantages of ABS Plastic

Advantages	Disadvantages
Impact resistant	Not well-suited for direct tooling
High tensile strength	Health hazards (carcinogen)
Heat resistant	Nonbiodegradable
Chemical resistant	
Scratch resistant	

In addition to ABS, Stratasys also uses the following materials:

- ABS color+
- Polylactate acid (PLA)
- Veroclear (clear filament)
- Polypropylene

The original Stratasys machines have a heated head (Figure 6.3) that extrudes the breakaway support system. As with the ABS plastic, this material is deposited as layers on the *X-Y* plane, and serves to support any overhanging portions of the prototype's geometry. As the name implies, the support material is easily broken away from the model.

6.2.3 The FDM Process

The FDM process forms 3D objects from CAD-generated solid or surface models. A temperature-controlled head extrudes thermoplastic material layer by layer. The designed object emerges as a solid 3D part without the need for tooling.

The process begins with the design of a geometric model on a CAD workstation. The design is imported into Stratasys' easy-to-use software, Insight, which mathematically slices the STL file into horizontal layers. The Supportworks software, also included, automatically generates supports if needed. The operator creates toolpaths with the touch of a button. The system operates in the *X-*, *Y-* and *Z*-axes, in effect, drawing the model one layer at a time.

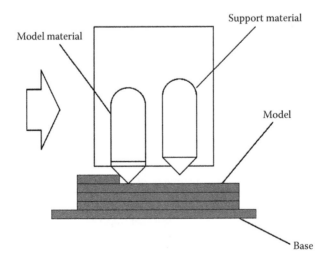

FIGURE 6.3
FDM heads.

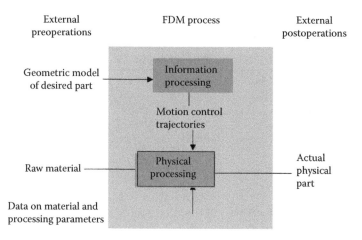

FIGURE 6.4
The FDM process.

Thermoplastic modeling material, 0.070 in. (0.178 cm) in diameter, feeds into the temperature-controlled FDM extrusion head, where it is heated to a semi-liquid state. The head extrudes and deposits the material in ultra-thin layers onto a fixtureless base. The head directs the material into place with precision and as each layer is extruded, it bonds to the previous layer and solidifies. A flow diagram of the FDM process is shown in Figure 6.4.

FDM systems are modular—allowing customers to increase system capabilities as their prototyping needs increase in complexity. The basic system includes all the necessary components to generate models and prototypes in one material. Modules are available to increase capabilities with a variety of model-making materials.

6.2.4 Machines

Over the years, Stratasys Inc. has developed many different products to facilitate different needs. Table 6.2 shows the technical specifications of three solid-based RP systems from Stratasys. A brief description of each of the products is given below.

6.2.5 Stratasys J-750

The Stratasys J-750 is the most precise of the Stratasys RP machines, making it useful when modeling fine-tuned objects. The J-750 also features the ability to prototype in multiple colors using its dual PolyJet color mapping system. The machine in Figure 6.5 is a medium-sized machine with a large print volume, making it very popular in the RP community [3].

TABLE 6.2

Solid-Based RP Systems by Stratasys

Machine	J-750	Dimension Elite	Objet Eden260VS
Work volume	19 in. × 15 in. × 7.5 in.	8 in. × 8 in. × 12 in.	10.0 in. × 9.9 in. × 7.9 in.
Layer thickness	14 μ (0.00055 in.)	0.01 in. (ABS color+) or 0.007 in. (ABS)	16 μ (0.0006 in.)
Accuracy precision	±0.0005 in.	±0.005 in.	±0.0006 in.
Material supply	ABS and PLA	ABS	ABS, Veroclear, polypropylene
Software	PolyJet Studio	CatalystEX	PolyJet Studio
Size	26.4 in. × 46.1 in. × 25.2 in.	27 in. × 36 in. × 41 in.	34.3 in. × 28.9 in. × 47.2 in.
Weight	335 lbs	282 lbs	582 lbs

FIGURE 6.5
The Stratasys J-750 rapid prototyping machine.

6.2.6 Stratasys Dimension Elite

The Dimension Elite is the smallest and lightest RP machine, making it useful for office or personal use. The CatalystEX operating system is an easy interface as well, making it convenient for those just getting involved in the RP community. The Dimension Elite does not have the highest precision, but is still able to prototype accurate and functional models and parts. It also has

FIGURE 6.6
The Stratasys Dimension Elite.

the ability to use ABS color+ material, meaning it can prototype in a larger array of colors than most other machines. This makes it a great machine for prototyping objects that require precise color mapping (Figure 6.6) [3].

6.2.7 Stratasys Objet Eden260VS

The Objet Eden 260VS is a RP machine useful when needing to use a variety of material types. It is able to prototype in ABS, Veroclear, and polypropylene materials, which is a greater range of material types than most other machines. However, the Eden260VS is a very large and heavy machine, meaning it needs a large storage space. Despite this, it still features a decent print area and precision, making it a viable and useful RP machine (Figure 6.7) [3].

6.2.8 FDM 3D Printing

FDM is the most common form of 3D printing. FDM printers use a solid plastic material to print objects. The two main types of materials are PLA or ABS filament. This filament is melted down inside the printer, then extruded into a shape, and then rapidly cooled and hardened to form the desired object. A variety of different companies use FDM printers, but the largest company is MakerBot, which is owned by the Stratasys corporation, but acts as a separate entity. The most popular of the MakerBot 3D printers is the Replicator.

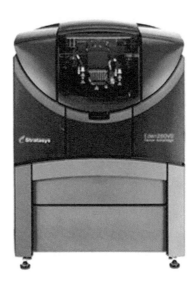

FIGURE 6.7
The Stratasys Objet Eden260VS.

6.2.9 MakerBot Replicator

The Replicator is a desktop 3D printer that is designed to sit on a table. The printer's build volume is 25.2 cm × 19.9 cm × 15.0 cm (9.9 in × 7.8 in × 5.9 in.) meaning it can make any part that is smaller than these dimensions and has a smaller volume than 7522 cm^3. The appearance of the printer and its dimensions can be seen in Figure 6.8. It can print items down to a precision of 100 μ (0.1 mm), making it capable of doing relatively fine details. At Loyola Marymount, the filament used is PLA filament. The Replicator uses this filament to rapidly create objects that are uploaded to the printer using a CAD model in STL format. A typical print takes from 3–13 h to print. The MakerBot Replicator costs $2,899, making it a reasonably priced FDM printer capable of effectively printing objects quickly with reliable precision.

6.3 SLA Systems

The original developer of SLA RP machines was 3D Systems, Inc. They designed apparatuses to produce solid, 3D plastic parts directly from CAD/CAM data. The company has quickly grown from pioneer to proven leader in RP with hundreds of installations worldwide.

Based in Valencia, California, 3D Systems, founded by Charles Hull, the inventor of SLA serves customers virtually worldwide, with offices in the

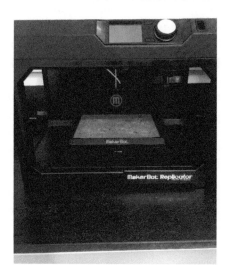

FIGURE 6.8
The MakerBot Replicator at Loyola Marymount.

United States, United Kingdom, France, Italy, Hong Kong, and Japan and employs more than 300 people worldwide.

6.3.1 The Details of SLA Process

The SLA process is fundamentally based on parts that are built from a photo-curable liquid resin that solidifies when sufficiently exposed to a laser beam (basically undergoing a photopolymerization process) that scans across the surface of the resin. The building is done layer by layer, each layer being scanned by the optical scanning system and controlled by an elevation mechanism which lowers at the completion of each layer. The details that SLA incorporates, such as the SLA process, the STL file format, C-Slice processing to STL files, thermoplastic resin, the SL manufacturing process, and quantitative analysis of laser-based manufacturing are described in this section [4].

6.3.2 The SLA Process

The stereolithographic process generates a part in the following manner:

1. A photosensitive polymer solidifies when exposed to an ultraviolet (UV) light source is maintained in a liquid state.
2. A platform that can be elevated is located just one layer of thickness below the surface of the liquid polymer.

3. The UV laser scans the polymer layer above the platform to solidify the polymer and gives it the shape of the corresponding cross-section. This step starts with the bottom cross-section of the part.
4. The platform is lowered into the polymer bath to one layer thickness to allow the liquid polymer to be swept over the part to begin the next layer.
5. The process is repeated until the top layer of the part is generated.
6. Postcuring is performed to solidify the part completely. This is required because some liquid regions can remain in each layer. Because the laser beam has finite size, the scanning on each layer is analogous to filling a shape with a fine color pen.

The SLA process of generating a part, described in the steps above, is illustrated in Figure 6.9. A picture of an SLA machine is shown in Figure 6.10.

6.3.3 SLA 3D Printing

SLA printers differ greatly in their print process from FDM printers. SLA printers typically use a thermal-plastic resin as their print material, which hardens into the model shape when a UV laser is shot into the resin. Although this is quite different from FDM printers, SLA printers also create objects layer by layer based on the CAD models uploaded to the printer. The Formlabs 1+ is a common SLA printer. The print volume for the Form 1+ is 125 mm × 125 mm × 165 mm (4.9 in. × 4.9 in. × 6.5 in.), meaning that it has a smaller print volume than that of the MakerBot Replicator [5]. The volume of the print area can be seen in Figure 6.11. Making up for the smaller print volume, the Form 1+ has a precision of 25 μ (0.025 mm), meaning it is four times more accurate than the MakerBot Replicator. However, the printer does take 6–16 h to print an object, which is longer than the MakerBot Replicator. The Formlabs 1+ costs $2,799, which is a very reasonable price for the precision and effectiveness of the printer.

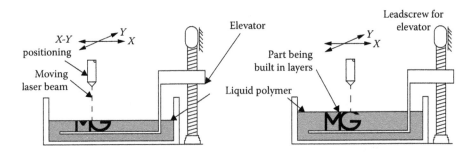

FIGURE 6.9
Stereolithographic process generating a part.

FIGURE 6.10
3D Systems SLA 3500 systems. (Courtesy of 3D Systems, Inc.)

FIGURE 6.11
The Formlabs 1+ at Loyola Marymount.

6.4 SLS Systems

6.4.1 SLS Overview

SLS process is introduced on a large scale by desktop manufacturing (DTM), using a laser to sinter a powdered material into the prototype shape. In many ways, SLS is similar to SLA except that the laser is used to sinter and fuse powder rather than photocure a polymeric liquid. In this process, a thin layer of thermoplastic powder is spread by a roller over the surface of a build cylinder and heated to just below its melting point by infrared heating panels at the side of the cylinder. Then a laser sinters and fuses the desired pattern of the first slice of the object in the powder. Next, this first fused slice descends one object layer, the roller spreads out another layer of powder, and the process continues until the part is built. As the device builds layer after layer, the unsintered powder acts as support for the part under construction [6].

The key advantage of SLS process is that it can make functional parts in essentially final materials. However, the system is complex mechanically than SLA and other technologies. Surface finish and part accuracy are not quite as good as those of other technologies, but material properties can be quite close to those of intrinsic materials. Since the prototyped parts are sintered, they are porous. Because of that, it may be necessary to infiltrate the part, especially metal parts, with another material to improve the mechanical properties. Figure 6.12 shows a simple SLS diagram.

The original SLS machines developed were RP machines created by Carl Deckard and Joseph Beaman at the University of Texas-Austin in the 1980s. Using powder-based material and a high-powered carbon dioxide laser, Deckard and Beaman were able to raise the powder to its melting point and fuse the powder particles together. By repeating this process layer by layer, the first SLS prototype model was successfully created. The original machine that accomplished this task was limited to only polycarbonate as its

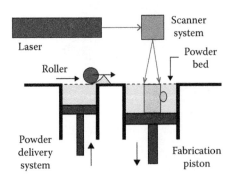

FIGURE 6.12
SLS diagram.

material, but modern RP SLS machines are capable of printing in polyvinyl chloride (PVC), ABS filament, nylon resin, polyester, and most importantly metallic powder.

6.4.2 3D Systems sPro 60 HD

An example of such a modern machine is the 3D Systems sPro 60 HD RP machine. This machine has a 15 in. × 13 in. × 18 in build envelope that can be precise to a level of 100 μ. The appearance of the sPro 60 can be seen in Figure 6.13. The sPro 60 utilizes a 30 W CO_2 laser that can move at 5 m/s, meaning it can make prototypes rather quickly (1.5 h build rate). This machine can use a variety of materials including basic metal powders, but works best with PVC or polycarbonate as its primary material. This makes it very useful as a modern day SLS RP machine.

6.4.3 Sinterit Lisa SLS 3D Printer

The alternative to the larger and more expensive SLS RP machines are SLS 3D printers. An excellent example of a SLS 3D Printing machine is the Sinterit Lisa. The build volume for the Lisa is 11 cm × 15 cm × 13 cm (4.3 in. × 5.9 in. × 5.1 in.),

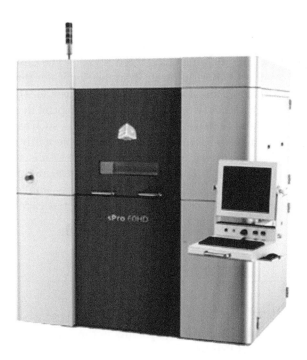

FIGURE 6.13
The 3D Systems sPro 60 HD RP machine.

FIGURE 6.14
The Sinterit Lisa SLS printer.

giving it the smallest build volume of the three printers. The Lisa can print to a precision of 60 μ, making it more accurate than the Replicator but less than the Form 1+. The appearance of the Lisa can be seen in Figure 6.14. The Sinterit Lisa sets itself apart from the other two printers in that it can print in multicolor in a single print and in its ability to use a variety of materials, such as Polyamide 12 (PA-12), a plastic-like material, reinforced nylon, and potentially metallic powder in the future. The average print time of the Sinterit Lisa is 4–14 h, making it middle ground in print time. The Lisa costs about $9889, making it the most expensive of the printers. Although the print volume is smaller and it costs much more, the Lisa is still an excellent printer through its capability of using a variety of materials and colors.

6.5 Thermal Inkjet Printing Systems

A final form of RP and 3D printing systems is the Thermal Inkjet style of printing. Thermal Inkjet Printing (TIJ) is a process that utilizes thermal, electromagnetic, or piezoelectric technology to place droplets of ink (sometimes actual ink or sometimes a melted alternate material) onto a powder substrate which hardens into a desired shape [7]. This process allows TIJ machines to be highly accurate (down to volumes of 10 pL), making it very useful in the biomedical field in areas like tissue engineering or regenerative medicine.

TIJ started with Siemens company in 1951 with the Rayleigh break-up ink-jet device. Initially, this technology was limited to the commercial industry, but in 2001, it began to be used in the medical industry. With the adoption of TIJ printing on a larger scale, more materials were being utilized, including polymer, ceramic, and glass-based materials. Currently, these materials are being improved to become more biologically compatible and useful in the medical industry [8].

6.5.1 Stratasys Polyjet Connex3 Inkjet 3D Printer

A TIJ printer that has become a leader in its field is the Stratasys Polyjet Connex3 printer. This printer has a build volume of 19.3 in. × 15.4 in. × 7.9 in., giving it a very large print volume compared to other 3D printers. An image of the Connex3 can be seen in Figure 6.15. This printer also has an accuracy of 16 μ, making it extremely precise. The printer can utilize all of the Stratasys's Vero materials as well as MED610, a material that has proven to be biologically compatible in tests so far. However, a typical print can take 8–20 h, making it the longest print time of any 3D printer. In addition, the

FIGURE 6.15
The Stratasys Connex3 Inkjet 3D printer.

TABLE 6.3

Comparison of the Different Types of 3D Printers (FDM, SLA, SLS, TIJ)

	MakerBot Replicator	Formlabs 1+	Sinterit Lisa	Stratasys Connex3
Pros	Fastest print time Large print volume Shortest orientation process	High precision Lowest cost Can create more complex shapes	Multicolor prints Use of multiple materials (durability of materials) Least amount of postprocessing	Highest precision Biomedical applications Largest print volume
Cons	Least precise Cannot make parts with high shape complexity	Largest amount of postprocessing Long print time	High cost Smallest print volume	Longest print time Highest cost

printer normally costs around $330,000, meaning it is normally only used by stable 3D printing companies [9]. Despite this, if it is possible to afford that price, the Connex3 represents the top end of TIJ printers.

6.6 Comparisons between Printing Processes

In Table 6.3, a comparison between the MakerBot Replicator (FDM), Formlabs 1+ (SLA), the Sinterit Lisa (SLS), and the Stratasys Connex3 (TIJ) is shown.

Based upon the type of part one is trying to print, each type of printer is optimal. For rapidly printed everyday objects, FDM printers are the best while for precise, complex parts, SLA printers are optimal. If a part needs to be strong and durable but not necessary to be large, SLS printers are ideal, while if a part needs to be exact and used in biotechnology, then TIJ printers are the way to go. The same can be said for RP machines; FDM and SLA machines are best for generalized prototyping while SLS and TIJ are best for precise parts. However, rapid prototyping is primarily for mass manufacturing and larger objects, meaning they are less common than 3D printers. This does not mean they are not useful, but rather that they have a more specified function.

6.7 Questions

1. What is the best way to classify 3D printers?
2. Which type of rapid prototyping process was developed first?
3. Describe the FDM process with figures.

4. Describe the SLA process with figures.

5. Describe the SLS process with figures.

6. Describe the 3D printing process.

7. Develop a table of comparisons between FDM, SLA process in terms of (1) Operational principle, (2) Source of energy, (3) Work volume, (4) Materials, (5) Cost, (6) Advantages, and (7) Disadvantages.

References

1. C.K. Chua and K.F. Leong. *3D Printing and Additive Manufacturing: Principles and Applications, Nanyang Technological University, Singapore.*

2. S. Crump. The extrusion of fused deposition modeling, *Proceedings of the 3rd International Conference Rapid Prototyping,* Eden Prairie, MN, pp. 91–100, 1992.

3. 3D Printing Solutions Stratasys. Stratasys.com, October 12, 2016.

4. R. Noorani, *Rapid Prototyping Principles and Applications,* John Wiley & Sons, Hoboken, NJ, 2006.

5. Replicator Desktop 3D Printer, Makerbot Industries, January 10, 2014. http://store.makerbot.com/replicator. Accessed June 28, 2016.

6. 3D systems, http://www.3dsystems.com, 2002.

7. Form 1+ High Resolution 3D Printer. FormLabs Online. June 10, 2014. Accessed June 28, 2016. http://formlabs.com/products/3d-printers/form-1-plus.

8. Sinterit Lisa 3D Printer. Sinterit Inc. November 18, 2015. Accessed June 30, 2016. http://sinterit.com/#home.

9. Connex3—Rapid tooling and prototyping in multiple materials | Stratasys. 2016. Stratasys.Com. Accessed October 30 2016.

7

Scanning and Reverse Engineering

7.1 Introduction

Reverse engineering (RE) refers to the process that creates a CAD model by acquiring the geometric data of an existing part using a 3D measuring device. It is widely used in various areas such as rapid product development, casting, NC machining, entertainment, part inspection, medical imaging, etc. This is because using RE can drastically reduce the development time and the cost of new products.

RE is a process that approaches the traditional engineering from the opposite direction. While traditional engineering turns engineering ideas into real parts, the RE process interprets the intended design idea in order to obtain a CAD model [1]. The ultimate goal of the RE process is, therefore, to obtain a computerized CAD model by digitally reconstructing an existing physical part.

Once a CAD model is built using RE techniques, the advantages of a CAD/CAM system can be fully appreciated. The lack of an existing CAD model may be attributed to various reasons, all of which require the use of the RE process. These are typical uses of RE where the CAD model for an existing part is not available [2]:

- A clay model is built by a designer.
- A part has gone through many design revisions without documentation.
- The drawing of a part is lost or no longer available.
- Digital preservation of historical artifacts or relics.

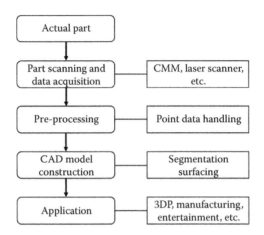

FIGURE 7.1
Conventional use of RE procedure.

In the conventional RE process shown in Figure 7.1, the CAD model is created based on point data collected from the part's surface by scanning or probing. This point data generally requires preprocessing operations such as removal of spikes and outliers, and point data reduction and arrangement. For CAD model construction, the well-organized point data are segmented for a curve-based surface modeling approach or meshed when a mesh-based surfacing process is used. After the point data have been captured using a 3D measuring device and the surfaces of the desired object have been reproduced, the file can be saved in various CAD formats. In particular, the stereolithography (STL) file format is preferred when it is desired to manufacture the part by rapid prototyping (RP) techniques. The computerized model produced by RE plays a very important role in many engineering and science applications. In this chapter, popular 3D measurement devices and principles, 3D model generation, and applications are presented.

The principles and applications of RE by 3D printing will be described here in this chapter. Topics such as scanning devices, model construction techniques, data handling and reduction methods, and applications and future trends will also be discussed.

7.2 Measuring Devices

7.2.1 Overview

RE is playing an increasingly important role in modern manufacturing. In the RE process, the CAD model of a product is generated from the measured point data of a physical prototype. In order to capture the shape of products,

contact- and noncontact-type measuring devices [3] are used in the industry as shown in Figure 7.2. When choosing a type of measuring device, it is
important to consider the part's shape, required accuracy, part material, size
of the part, and captured data type.

Contact- and noncontact-type measuring devices are the two types
of devices used in sampling point data from a part's surface. A common
contact-type measuring device is the coordinate measuring machine (CMM),
which has been used extensively for inspection and tolerance. This type of
machine is usually NC-driven and can obtain point data with an accuracy
of several microns. However, it is inherently slow in acquiring point data as
it needs to make physical contact with the part's surface for every point that
is sampled. Due to the shortcomings of CMMs, it is inefficient and difficult
to measure a freeform part. Therefore, CMMs are used for automatic inspection of simple primitive shapes such as slots, steps, holes, and pockets.

Unlike contact-type devices, noncontact-type devices do not require physical contact with a part to capture point data. For this reason, noncontact-type
devices have become a more viable option for capturing large amounts of
point data very quickly. Among various noncontact-type measuring devices,
the laser scanner is one of the most dominant systems. This is due to the
fact that laser scanning devices can acquire a large amount of point data in
a short period of time with a relatively high degree of accuracy. As the accuracy of laser scanning devices has improved, the use of these devices in the
industry has expanded significantly. In addition to the laser scanners, computed tomography (CT) scanners are used for the scanning of the internal
shape and material properties of parts. From medical CT data, the human
body can be visualized, and the data can be used to make implants for surgery. For industrial products, the internal shape, porosity, and density information can be safely acquired using high-power industrial CT scanners.

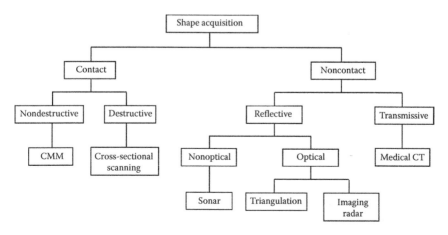

FIGURE 7.2
Classification of measuring devices.

TABLE 7.1

Characteristics of Contact- and Noncontact-Type Measuring Devices

	Major Influence Factors for Measuring	Pros and Cons	Shapes
Noncontact devices	• Lighting condition • Surface roughness • Reflectance characteristics • Scanning direction • System calibration • CCD resolution, etc.	• Fast measuring speed • Applicable to freeform surfaces • Applicable to soft materials • Relatively low accuracy • Large number of point data	• Freeform surfaces • Boundary edges • Casted/molded objects
Contact devices	• Material • Approach direction/speed • Probe contact point • Surface roughness • Size of details • Number of sampling points • System calibration • Stylus length • Accessibility	• High accuracy • Can measure transparent part • Relatively slow measuring speed • Various application S/W available	• Primitives • Deep holes • Machined parts

Table 7.1 summarizes some of the major influencing factors for contact and noncontact devices and the major pros and cons of each [4].

7.2.2 Contact-Type Measuring Devices

As previously stated, the CMM has been widely used in industry for inspection and RE of products. Although it is highly accurate, the speed of capturing surface point data is extremely slow. Therefore, CMMs are mainly used for measuring parts with well-defined primitive shapes such as slots, steps, holes, and pockets. Figure 7.3 shows an example of a commercial touch probe. Traditional probes consist of a probe head, part extension, probe, and stylus. There are many variations of touch probes in terms of the type of probe head, the type of probe, the length and type of stylus, and tip diameter.

Each point that is measured on a particular workpiece is unique to the machine's coordinate system. One can obtain the point data of the part by controlling the touch probe manually or automatically in Figure 7.3 [5]. The accuracy of the captured point data depends on such variables as the structure of hardware, control system, approach direction, approach speed, and atmospheric temperature. Therefore, it is important to measure a part with the proper hardware in a well-controlled environment.

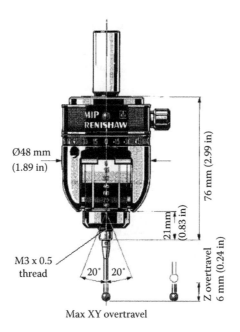

Max XY overtravel

FIGURE 7.3
An example of a touch probe.

7.2.3 Noncontact-Type Measuring Devices

7.2.3.1 Laser Scanning Systems

Laser scanners are noncontact-type measuring devices that use active scanning methods to capture the 3D shape of a part. The laser probe emits a laser beam to the part, and the CCD cameras then capture the 2D image of the projected beam [6]. From the captured images and the structure of the hardware system, the 3D coordinates of a part can be obtained by applying triangulation techniques. A laser probe is usually mounted on a multiaxis CNC-type mechanism or on the end effector of a robotic arm.

When using laser scanners to reverse engineer a desired shape, it is important to remember that there are several constraints that must be satisfied to produce quality scans. These constraints include the view angle, field of view, depth of field, incident beam interface, and the travel path. Other factors that can influence the quality of scans are the roughness and reflectiveness of the object's surface. Figure 7.4 illustrated a basic laser scanning mechanism.

Laser scanning devices can have a range of costs depending on the size and accuracy desired. For example, the NextEngine Desktop scanner (Figure 7.5) has a relatively low cost (<$5000), but is tailored to smaller and less complex parts. This type of scanner serves well in a classroom or office setting because it is small and portable as compared to other more expensive laser scanners.

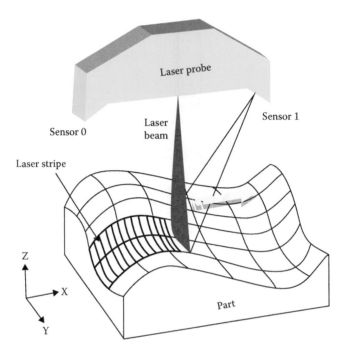

FIGURE 7.4
Laser scanning mechanism.

FIGURE 7.5
NextEngine 3D desktop scanner.

It is ideal to use this type of scanning device to reverse engineer objects such as small mechanical parts (brackets, face plates, and clamps). The NextEngine scanner is a good place to start for students and professionals looking to gain RE experience. Case studies performed using the NextEngine Desktop scanner can be found at the conclusion of this chapter.

Another much more expensive scanner is the Artec Eva sold for nearly $20,000. The Artec Eva is a handheld scanner which captures over 16 frames per second to automatically align in real-time (Figure 7.6). This unique feature allows the Artec Eva to scan objects extremely quickly with very little effort. Using the Artec software, multiple scans can be aligned together annually to fill any gaps or holes in the initial scan and create a single 3D model. In industry, the Artec Eva scanner is primarily used for scanning one's face and body, for animation and computer graphics, and for capturing color and texture of an object. While this scanner captures objects very quickly and easily, it has lower data density and accuracy than other much slower scanners. It is recommended that this scanner be used when portability and speed are the most important factors as opposed to accuracy.

Using blue light technology and a 5 megapixel camera to capture images, the Steinbichler Comet L3D is an extremely accurate noncontact scanning device (Figure 7.7). This 3D scanner uses LED lighting in order to ensure excellent results despite poor ambient conditions. This is achieved by an LED pulse mode and a relatively short working distance. Due to these unique features, the Comet L3D produces highly dense, accurate, and repeatable scans. In industry, the Comet is used in a range of applications including quality control and inspection, mold and tool making, rapid manufacturing, and archeology.

The final scanner that will be explored in this section is the Surphaser 100HSX 3D laser scanner (Figure 7.8). This scanner is best known for its sub-millimeter accuracy, ability to capture up to 1 million points per second, and

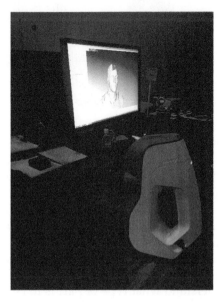

FIGURE 7.6
Artec Eva scanning device.

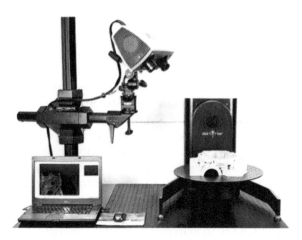

FIGURE 7.7
Comet L3D scanner.

FIGURE 7.8
Surphaser 100HSX scanner.

a scanning range of 1–50 m. This scanner features an optional 60 megapixel camera for capturing color images with automatic color data mapping. The Surphaser is primarily used in the aerospace, automotive, structural, and tooling industries due to its highly dense and accurate parts that can cover a large scanning volume.

There is a wide range of RE scanners that one can choose depending on the cost, speed, accuracy, and size of the part desired. When selecting a scanner, it is important to understand what type of part is to be recreated and what aspects of RE are most important. Table 7.2 below compares the four scanners discussed in this chapter.

TABLE 7.2

Comparison of Scanning Devices

	NextEngine	Eva Artec	Comet L3D	100HSX
Cost	<$5,000	~$20,000	~$80,000	~$100,000
Advantages	Portable	Portable	High density	High density
	Fast	Fast	Accuracy	Accuracy
	Low cost	Medium cost	Repeatability	Large scan volume
	User friendly		Variable volume	
Disadvantages	Poor accuracy	Low-to-medium	Long scan time	Long scan time
	Low-density	accuracy	required	
	scan			
Applications	Small mechanical	Facial scanning	Aerospace/	Aerospace/
	parts		automative	automative
	Color/texture	Animation	Tooling	Tooling
		Color/texture	Inspection	Environment/
				scenes
				Structural

7.2.3.1.1 Example of Laser Scanning

As an example, Figure 7.9 shows the laser scanning of an eardrum model. Figure 7.9a shows a plastic model of an eardrum. The model is attached to the motorized stage and scanned using a laser scanner (Surveyor 1200 by LDI, Inc.). After scanning, the point data captured in each direction are registered on one coordinate system (Figure 7.9b). Finally, in Figure 7.9c, the scanned point cloud is used for the generation of computer-aided engineering mesh data for acoustic analysis [7].

7.2.3.2 CT Scanning Systems

CT is a nondestructive evaluation tool for measuring interior shapes, density, and porosity of products. CT scanners do not require elaborate fixturing, positioning, or part specific programming and can generate dense, well-behaved clouds of coordinate data. Points extracted from CT

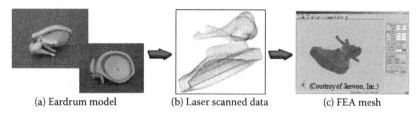

(a) Eardrum model   (b) Laser scanned data   (c) FEA mesh

FIGURE 7.9

Laser scanning of an eardrum model. (a) Eardrum model. (b) Laser scanned data. (c) FEA mesh.

images have known connectivity, surface topology, and surface normals. CT scanners have been widely used for medical applications, and in recent years, industrial CTs have been developed and utilized for manufacturing applications [8].

7.2.3.2.1 Medical CT Scanners

In the past, medical doctors diagnosed patients using 2D X-ray images based on their experience. They imagined a 3D human body for surgery from a series of X-ray images. This made it difficult for doctors to make an accurate assessment of the patient's condition, and in the case of surgery, almost impossible to design perfectly compatible implants. By using CT scanners, doctors have been able to visualize a patient's specific anatomical characteristics before surgery and then generate data for machining and analysis of implants from the CT data. In a typical medical application, one-of-a-kind parts are needed for each patient and these parts must be manufactured not by mass production but by order-adaptive production. Through an RP process with medical grade materials, these parts can be manufactured in a short time using CAD models from medical images. Figure 7.10 shows the principle of spiral CT scanning, and Figure 7.11 shows a medical CT scanner [9].

7.2.3.2.2 Industrial CT Scanners

Industrial CT scanners are nondestructive inspection devices that have had a significant impact in aeronautical and space applications. Due to recent advances in the industrial CT technology, especially in terms of data processing methods, the field of industrial CT is expanding into several new areas of application. Data processing methods are currently available that allow for (1) CT-assisted RE, (2) CT-assisted metrology, and (3) CT-based finite-element analysis. CT-assisted RE is the process of going from CT data directly to a CAD model. CT-assisted metrology is the process of extracting geometric data from a component to compare with the original design

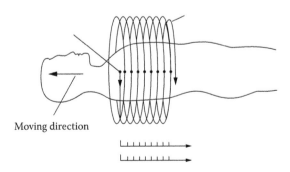

Moving direction

FIGURE 7.10
Principle of spiral CT scanning.

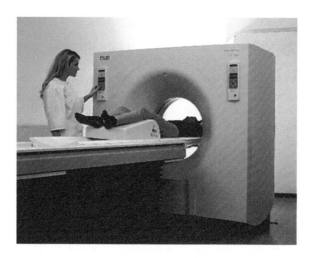

FIGURE 7.11
Medical CT scanner.

(a) Industrial scanner   (b) CT images of automobile engine

FIGURE 7.12
(a) Industrial scanner. (b) CT images of an automobile engine.

dimensions. Figure 7.12a shows a commercial CT scanner used in Jesse Garant Metrology Center. An example of an industrial 3D scan generated from a CT image is shown in Figure 7.12b at the Jesse Garant Metrology Center.

7.3 CAD Model Construction from Point Clouds

Two major approaches can be used to generate 3D models from point clouds: the curve-based and the polygon-based. However, it is necessary for the point cloud data to be correctly preprocessed before building a 3D model.

7.3.1 Preprocessing

Preprocessing operations in RE include *registration, noise removal*, and *data reduction* steps.

In the case of a complex shape, the scanner cannot capture the complete surface data from a single scan direction. Thus, the object must be scanned multiple times, with a variation in the object orientation. Every time the object is scanned, a portion of the total point cloud is obtained. Therefore, it is necessary to combine all the fragmented point clouds into one single coordinate system. This procedure is usually referred to as *registration* and is shown in Figure 7.13.

In data acquisition, point clouds often contain outlier points or shadow points that complicate further computation and processing. *Noise removal* uses computation based on the distribution of points across a point cloud in order to eliminate unnecessary outliers from the dataset. This is an important step in creating a smooth and accurate surface model often represented by a Nonuniform Rational B-Spline (NURBS) model. Common noise removal methods in RE include *Gaussian, average*, and *median* methods. All of these methods are based on statistical approximation.

With recent advances in scanning devices, a large number of point clouds can be gathered in a relatively short time. These highly dense point clouds, a number which could easily reach several million data points, provide a burden for further processing. Such an overwhelming amount of points can contain a significant number of redundant points and can require much more computation time. Therefore, a *data reduction* procedure is needed to reduce the number of point clouds while maintaining the accuracy as illustrated in Figure 7.14. The key idea in data reduction is to

- Remove the points on flat regions.
- Keep the points on highly sculptured regions such as boundaries, internal edges, and corners.

7.3.2 Point Clouds to Surface Model Creation

Surface creation plays an important role in the RE process. Since complex-shaped objects are usually the targets for RE, an efficient method is crucial to create accurate surface models in a short time. Compared to conventional

(a) Scheme of registration process   (b) Registration using artificial features

FIGURE 7.13
Registration. (a) Scheme of registration process. (b) Registration using artificial features.

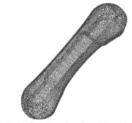 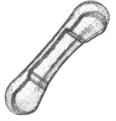

(a) Original point clouds with redundancy   (b) The reduced point clouds

FIGURE 7.14

Data reduction example (phone model). (a) Original point clouds with redundancy. (b) The reduced point clouds.

surface modeling methods, RE puts a greater emphasis on the creation of *free-form* surfaces. The methods of converting point clouds to freeform surfaces can be categorized into the two main approaches: curve-based modeling and polygon-based modeling, which are summarized in Table 7.3 shown below.

In the curve-based modeling shown in Figure 7.15, the scanned point clouds are first rearranged into a regular pattern, usually a series of cross-sections. These point clouds are then subdivided or classified into simpler shapes of point sets. Once we have classified all subdivided point clouds, the surface fitting can be performed. Generally, surface fitting begins with curve fitting. The cross-sections of point clouds are fitted with NURBS curves using the least squares method. With the skinning function, the fitted curves are approximated to a NURBS surface model. Finally, each skinned surface is combined to make a valid CAD model.

Most current commercial modeling systems such as CATIA, ICEM/SURF, IMAGEWARE, SOFTIMAGE 3D, and ALIAS/WAVEFRONT can be classified into curve-based modeling systems. These systems have made it possible

TABLE 7.3

Comparison of Curve-Based Modeling vs. Polygon-Based Modeling

Curve-Based Modeling	Polygon-Based Modeling
Segmentation, curve fitting, and skinning	Triangulation, decimation, subdivision, and triangle to NURBS fitting
Manual processes (a few weeks to a few months)	Automatic processes (a few hours to a few days)
Manual surface patching	Automatic surface patching
High-quality surfaces	Medium or Low-quality surfaces
Surface geometry only	Surface geometry and attributes (e.g., color)
Cannot ensure continuity	Ensure continuity
Simple topology	Complex topology

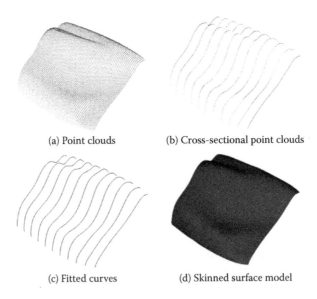

(a) Point clouds   (b) Cross-sectional point clouds

(c) Fitted curves   (d) Skinned surface model

FIGURE 7.15
Curve-based modeling. (a) Point clouds. (b) Cross-sectional point clouds. (c) Fitted curves. (d) Skinned surface model.

to create CLASS A (high-quality) surface models. However, it is difficult to create complex-shaped surfaces using this approach. For instance, if an engineer wants to design a car body panel using this approach, it often takes a few weeks to months to construct a quality model.

On the other hand, the NURBS surfaces can be directly created from triangular data without fitting curves. The main characteristic of the polygon-based modeling is to construct a polyhedral model before making a NURBS model. The construction of a polyhedral model is performed by connecting neighboring points based on Delaunay triangulation, alpha shape, crust, or volumetric method [10]. The triangulated data are then refined by the decimation and subdivision algorithms. These methods have been developed recently in the field of computer graphics and would produce adequate quality data to use in application areas such as entertainment, RP, and web visualization. If the application requires a surface model, the NURBS fitting from the triangulated data must be performed.

Some software commercially available for this polygon-based modeling includes GEOMAGIC, RAPIDFORM, and PARAFORM. Compared to the curve-based modeling systems, this procedure can construct a NURBS model in a few hours to a few days regardless of the geometric complexity of the scanned parts. Moreover, the recently developed scanning devices can capture surface attributes such as color and texture in addition to the geometry. However, it is still difficult to reconstruct fine details such as sharp corners and sharp edges with this approach.

Key algorithms used for polygon-based modeling include decimation, subdivision, and triangles to NURBS fitting. In decimation, the goal is to reduce the number of triangles in the polygon model while minimizing the approximation error. The most popular way to achieve decimation is by edge collapse which calculates the cost of each edge by estimating the error that occurs when it is removed. An example of edge collapse is illustrated in Figure 7.16.

On the other hand, subdivision seeks to generate a smooth and fine surface from the coarse mesh. By applying a refinement rule, we can obtain a smooth and detailed model. Subdivision surface schemes allow us to take the original polygon model and produce an approximated surface by adding vertices and subdividing existing polygons. There are many existing subdivision schemes. An example of a subdivision surface is shown in Figure 7.17. In this case, each triangle in the original mesh on the left is split into four new triangles [11].

Finally, triangle to NURBS fitting is an algorithm that constructs the initial parametric domain over a given set of triangles and approximates the tensor B-spline patch with continuity. Adaptive refinement is then performed in order to achieve the required accuracy of the fitted surface. Figure 7.18 shows an example of this method.

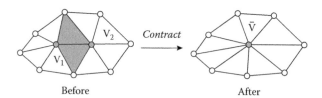

Before   After

FIGURE 7.16
Edge collapse.

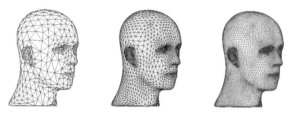

FIGURE 7.17
Example of subdivision surface.

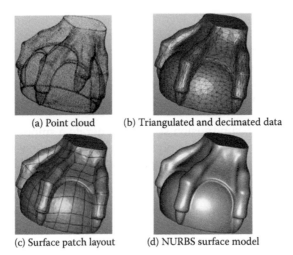

(a) Point cloud   (b) Triangulated and decimated data

(c) Surface patch layout   (d) NURBS surface model

FIGURE 7.18
Example of conversion of triangles to a surface model. (a) Point cloud. (b) Triangulated and decimated data. (c) Surface patch layout. (d) NURBS surface model.

7.3.3 Medical Data Processing

Medical images are generally represented by a voxel dataset, which is constructed by an arbitrary number of slice images with the same scanning intervals. The sample points of the volume data are called the voxel (volume element) since they are spatial objects. There are two main schemes to reconstruct 3D models from the volume data: *contour extraction* and *triangle extraction*. In contour extraction, each slice is processed in order to obtain contours using image processing. The extracted contours are further converted to a NURBS model by skinning. In triangle extraction method, a set of triangles is extracted between consecutive slices by the triangle extraction. This approach is useful in 3DP applications since no additional procedures are required to generate a file with the STL format.

7.4 Data Handling and Reduction Methods

Data handling and reduction have become important topics in RE. While many data handling and reduction methods have been proposed for image processing, most have been designed for dealing with the meshed point data. Only a few methods have been developed that can be directly applied to the point data generated from measurement devices. There are two methods that are used for data handling and reduction when it comes to point data generated from measurement devices. These two methods are the *uniform grid method* and *nonuniform grid method*. The uniform grid method creates a grid plane consisting of equivalent-sized grids perpendicular to the scanning direction. One point is then selected from within each grid based on median-filtering rule to be extracted. This method is good for maintaining the original point data set as it selects points rather than changing their positions. It is especially useful in cases where data reduction needs to be done very quickly for parts with relatively simple surfaces (see Figure 7.19).

In the case of applying the uniform grid method, some points for which the part shape drastically changes, such as edges, can be lost because no consideration of part shape is provided. In RE, it is critical to accurately recreate part shape, and the uniform grid method has limitations in this regard. In cases where the limitations of uniform grid method present too many challenges to RE a part, nonuniform grid method is used instead. There are two levels of nonuniform grid method: unidirectional and bidirectional. They can be applied considering the characteristics of the measured data.

Unidirectional nonuniform grid method is primarily recommended for when a part consisting of simple surfaces has more points along one direction (v-direction) that need to be measured as compared to the other direction (u-direction). On the other hand, when a part to be measured has complex and freeform surfaces, the point data is supposed to be dense along both the u and v directions. In this case, the bidirectional nonuniform grid method is more appropriate than the unidirectional nonuniform grid method.

7.4.1 3D Grid Methods

The 3D-grid method can handle the entire surface of a 3D object whether it is a single point cloud or multiple point clouds [12]. In the case of multiple point clouds for a single object, they need to be registered under one coordinate system. The proposed method uses the normal values of points on the

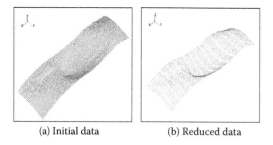

(a) Initial data   (b) Reduced data

FIGURE 7.19
Result of uniform grid method for free-form shape. (a) Initial data. (b) Reduced data.

part surface, from which the 3D nonuniform grids are generated using the standard deviation of normal values. Data reduction is performed by selecting one representative point and discarding the other points from each grid.

7.4.1.1 *Extraction of Points*

As a result of grid subdivision, many grids are generated where the part geometry drastically changes, whereas a few grids appear where the part geometry shows little change. From these grids, the points that can represent part geometry are extracted. In selecting a point that represents the points within a grid, an average of normal values is used. Therefore, a point whose normal value is closest to the average of the points within a grid is chosen. The selected point is regarded as the most representative point among the points within the grid (see Figure 7.20d).

In this method, the level of data reduction is basically determined by two factors: the number of initial grids and the size of the user-defined tolerance. For surface fitting of a plane as an example, the data points need to appear within a certain interval so that a surface fitting operation can be performed. In this case, the size of the initial grids is determined by these intervals and the number of initial grids is decided accordingly.

In the developed program, the number of initial grids is determined before the tolerance. The tolerance is the main factor in data reduction; the smaller

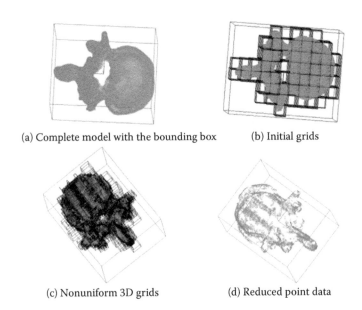

(a) Complete model with the bounding box   (b) Initial grids

(c) Nonuniform 3D grids   (d) Reduced point data

FIGURE 7.20
Example of a human backbone. (a) Complete model with the bounding box. (b) Initial grids. (c) Nonuniform 3D grids. (d) Reduced point data.

the tolerance the more points are left, and vice versa. If the user wants the remaining points to be distributed in certain grids to keep the accuracy, the smaller size initial grids must be used. If the user wants less number of points while keeping the accuracy, the reduction process should start with a lesser number of initial grids with a high tolerance.

7.4.1.2 Analysis

A phone model shown in Figure 7.21a was used for evaluating the performance of the 3D-grid method. In this case, the point data were simulated by converting a surface model into an STL model with a small tolerance. The nodes of the STL model were regarded as measured points and no noise was added to the point data so that they can be considered as point data that already had noise filtering.

Figure 7.21b through e shows the phone model data for different methods using data reduction of 90%. The phone model shown in Figure 7.21b by the 3D-grid method shows more points distributed at the edges compared to those shown in Figure 7.21c and Figure 7.21d by uniform or space sampling methods. For the point data reduced by chordal deviation sampling, the edges seem to be well-preserved as shown in Figure 7.21e, however, the remaining point data do not perform as well for surface fitting. When using the same number of points, the phone model generated by the 3D-grid method keeps better details at edges.

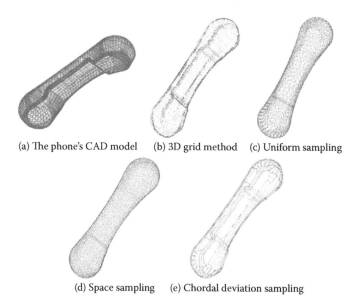

(a) The phone's CAD model   (b) 3D grid method   (c) Uniform sampling

(d) Space sampling   (e) Chordal deviation sampling

FIGURE 7.21
Reduced point data of the phone model. (a) The phone's CAD model. (b) 3D grid method. (c) Uniform sampling. (d) Space sampling. (e) Chordal deviation sampling.

7.5 Applications of RE

7.5.1 Applications

RE is the method of creating 3D computer models from physical parts for which no design data currently exists. As the technology of RE has advanced the number of applications in which RE can be applied has grown significantly. Among many of these applications, some of the most prominent are product development and manufacturing, entertainment, and biomedical engineering. The details of these areas are described in the following sections.

7.5.2 Product Development and Manufacturing

Product development and manufacturing is one of the most important areas where RE has been used. Applications in product development and manufacturing have included esthetic styling and refinement, tooling design and production, customer product prototypes, and part inspection.

RE is well-known as a key process in designing car bodies [13]. The esthetics of a car body are very difficult to draw directly by mathematical formulas in 2D or 3D. The solution is to create a CAD model from a clay model of the car body. Figure 7.22 shows the measured data and the CAD model created from the data. Similarly, CAD model generation from a mockup is one of the major processes in any styling job.

When creating a mold, manual modifications occur frequently, and the final shape of the mold often changes from the original design data. Therefore, the original CAD model can quickly become obsolete and a new CAD model for the modified mold must be obtained. Figure 7.23 illustrates an example of a mold that was the product of a CAD model. It displays a spoon shape in a mold that was created using the RE process.

Inspection occurs after using the mold and is the process that compares a manufactured part with its original CAD model to verify whether the

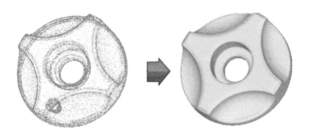

FIGURE 7.22
Data scan (left) and CAD model (right) of bolt part.

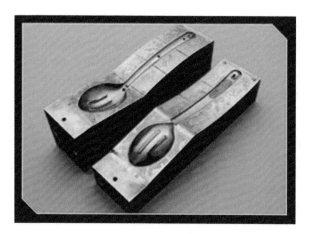

FIGURE 7.23
RE application in mold design.

part coincides with the original design data. The RE process can be directly applied to inspection. In inspection, either the measured point data are directly compared with the CAD model or the created surface model is compared with the CAD model.

Customization of products is one of the key trends in present markets. Clothes, helmets, personal products, and shoes are typical products that can benefit from customization. Figure 7.24 illustrates an example of a CAD-designed pocket knife that is customized for the potential user.

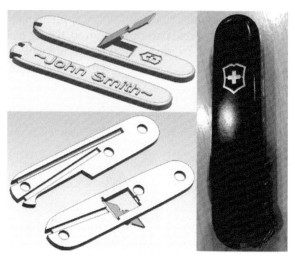

FIGURE 7.24
Design of customized product in the form of a pocket knife.

7.5.3 Entertainment

The desire for entertainment is rapidly increasing. Movies, games, and toys occupy a large part of children and adults' lives. For some movies, these days, animation and computer graphics often play an important role. Because human and animal animation models are very complicated, in many cases, they are created using RE techniques. Avatars used in 3D games are similar to the ones in the movies. Similarly, in the toy industry, mockups of characters are created first and then converted to CAD models. As observed, the parts that consist of freeform and complicated surfaces are targets of the RE process. Applications of RE in the entertainment industries are illustrated in Figure 7.25. It shows an animation character from *Geri's Game* which received the Academy Award for Best Animated Short Film in 1997 [14]. The control mesh of the human character was created by digitizing a full-scale model sculpted out of clay. The smooth (polygon) surfaces were created by subdivision techniques.

7.5.4 Biomedical Engineering

Biomedical engineering is one of the newest areas where RE has been partnered with RP techniques. Since human bones have complicated shapes, NC machining of these parts is difficult. RP techniques can be used to fabricate them, with a CAD model created by RE techniques. Measurement data for human bodies are acquired by CT/MRI scanners or noncontact measuring devices like laser scanners. Generated CAD models of bones can be used for many applications such as implant design and surgery planning. Figure 7.26 shows the CAD models of knee bones, skull, and teeth reconstructed by the RE process [15].

7.5.5 Other Applications

In addition to the applications described above, many other areas of application exist. Where digital models for real parts are required, the RE technique can be applied. Some example areas include cultural assets, archeology, web3D, and

(a) Control mesh model   (b) Model with rendering

FIGURE 7.25
Examples of animation (*Geri's Game*). (a) Control mesh model and (b) model with rendering.

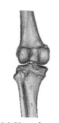 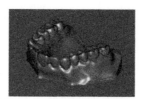

(a) Knee bones   (b) Skull   (c) Teeth (courtesy of geomagic, Inc.)

FIGURE 7.26
Biomedical application of RE. (a) Knee bones, (b) skull, and (c) teeth. (Courtesy of Geomagic, Inc.)

virtual environment. Interest and needs for developing web3D and virtual environment are increasing rapidly. With the help of RE techniques, any real parts can be shared as 3D models in websites or virtual environments. It is also essential in many E-commerce applications. In summary, RE is very efficient, has many potential applications, and provides an innovative design paradigm. By combining RE with RP technologies, the cycle for product development can be greatly reduced while the quality of the end product is enhanced.

7.6 Case Studies

7.6.1 Case 1: Recreation of Mechanical Parts

7.6.1.1 Introduction

The emergence of RE technologies and the rebirth of RP are fueling a business growth in all aspects of design, product development, and manufacturing. Companies engaged in product development and manufacturing are in tremendous competition to bring a product to the market faster, cheaper, with both higher quality and functionality. There are many situations where mechanical parts need to be produced but no design data currently exists. Thus, it was determined that RE techniques should be used to recreate a mechanical part. For this case study, a low-cost RE system, the NextEngine Desktop scanner, was used to obtain the geometric surface data of the part using 3D noncontact laser scanning and surface reconstruction techniques. The general strategies used to reconstruct the mechanical part using RE and RP are illustrated in Figure 7.27.

7.6.1.2 Methodology

The geometric data, in this case, was captured using noncontact-based laser scanning. The original mechanical part was a small steel muffler removed from a Honda Gxh50 four stroke engine. This was a relatively complex part

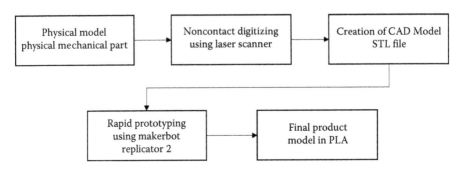

FIGURE 7.27
General strategies of the proposed approach.

whose geometry and reflective surface presented many challenges. Due to the small size of the part, a desktop 3D scanner was used to capture and reconstruct the surface geometry. The scanner initially took 2D images of the object before a stream of laser beams were swept across the object to record millions of data points along the object's surface. The millions of data points collected by the scanner were used to produce a point cloud from which a valid STL model could be derived (Figure 7.28).

The complex geometry and reflective surface of the part resulted in many defects in the point cloud that made polygonization very difficult. Typical errors in the data were as follows:

FIGURE 7.28
Original muffler.

- Holes/gap
- Uneven surface transitions between stitched scans
- Obscured point data due to reflections off part surface

These errors were corrected using NextEngine's ScanStudio software and Netfabb 3D Printing software.

7.6.1.3 Prototype of Mechanical Part

The prototype was made using fused deposition modeling (FDM). It was built using a MakerBot Replicator 2 3D Printer with polylactic acid (PLA) plastic. Figure 7.29 shows the prototype of muffler in PLA plastic.

7.6.2 Case 2: RE Prosthetics

7.6.2.1 Introduction

Rapid Prototyping is a technology developed in the late 1980s in order to fabricate physical models of 3D computer-aided designs (CAD). Since that time, RP has faced many challenges along its way to becoming one of the today's most quickly expanding and versatile technologies. These challenges include dimensional accuracy, the speed of production, and materials available. As these challenges have been tackled by greater research and technological advancements, the full scope of RP applications has greatly expanded. Recently, RE has emerged as a useful tool to create models for which there is

FIGURE 7.29
Rapid prototype of RE muffler.

no design data. This has proven especially useful in the biomedical field in the creation of prosthetic limbs. There are many accident victims who have lost various limbs. While many prosthetics can cost as much as $30,000, a low-cost reverse engineered prosthetics only costs a few hundred dollars. This is giving hope to many who cannot afford the much more expensive models. For this reason, this case study focused on exploring the production of biomedical devices using low-cost reverse and RP engineering techniques.

7.6.2.2 Methodology

A noncontact-based laser scanning device was used to scan the mold of a chosen subject's hand to obtain the point cloud data. The data files obtained from the scans were used to create a computer model of the hand that could be converted to an STL and imported into a CAD modeling program. For this case study, SolidWorks was the chosen as CAD software. SolidWorks was used to further manipulate the hand model to make way for cables to control movement and a mount to attach the prosthetic to the user's wrist. The retro-fitted computer model was then used to rapid prototype the subject's fingers so that they may be constructed into a prosthetic. Once mechanical cables and velcro straps are added to the prosthetic physical prosthetic, the user will be able to perform common tasks such as picking up objects. The ability of the prosthetic to pick up small objects including a cup, a water bottle, and a pencil will be used to evaluate the effectiveness of the low-cost RE system to recreate human fingers for biomedical applications.

7.6.2.3 Mold of Human Fingers

The mold that was used to reverse engineer the human fingers was made from a plaster-based material that hardened within 3 days of being poured. The eventual prototype of the prosthetic will be built using FDM with PLA plastic. Figure 7.30 shows the plaster mold of the human thumb.

The mold allowed for a relatively smooth surface that could be easily scanned with only minimal challenges. The challenges that resulted from this case study were overcome using the ScanStudio software provided by NextEngine with no need for further manipulation.

7.6.2.4 Computer Model of Human Fingers

The computer model produced using the NextEngine Desktop 3D scanner showed smooth surface transitions and overall quality. Any hole/gaps pro-duced in the computer model were easily handled by NextEngine's intui-tive ScanStudio software. While minor flaws will almost always occur when using low-cost systems, the future physical prosthetic produced from this computer model will be extremely accurate. Figure 7.31 shows a computer model of the human thumb recreated by the NextEngine 3D scanner.

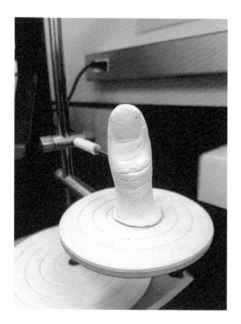

FIGURE 7.30
Thumb mold scanned.

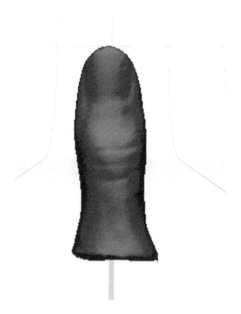

FIGURE 7.31
RE thumb in ScanStudio software.

7.7 Summary

In this chapter, the principles and applications of RE were introduced and discussed. The chapter began with an overview of measuring devices. Both contact and noncontact-types were discussed.

The construction of CAD models from point clouds is an essential component of RE. Both the curved-based and polygon-based approaches for CAD generation were described in detail.

Another issue of RE is data handling and reduction. While many data handling and reduction methods are available, 2D and 3D grid-band data reduction methods were described in this chapter. The uniform and nonuniform grid and 3D grid methods were also discussed.

The chapter then discussed significant applications of RE. Lastly, in order to better comprehend the principles of RE, several case studies were presented, showing step-by-step procedures that were taken for the reproduction of objects for which no design data existed.

7.8 Problems

1. Which type of measuring device is more suited to measure a freeform surface, contact type or noncontact type? And why?

2. What others can be measured by an industrial computed tomography device than the internal shape?

3. Explain the role of registration in the preprocessing of point clouds?

4. Laser scanning mechanism uses the triangulation method to get shape information from the physical objects. Describe the principle of triangulation in terms of x, y, and z coordinates of measured points.

5. What is the main difference and characteristics between curve-based modeling and polygon-based modeling?

6. The medical images such as MR images and CT images are described by voxels. (1) What is the main geometric parameters in the voxel representation? (2) What is the material property of the CT image? (3) Discuss possible applications of medical images in conjunction with the RE techniques.

References

1. T. Várady, R.R. Martin, and J. Cox, Reverse engineering of geometric models an introduction, *Computer-Aided Design*, Vol. 29, No. 4, pp. 255–268, 1997.
2. K.H. Lee, H. Woo, and T. Suk, Point data reduction using 3D grids, *International Journal of Advance Manufacturing Technology*, Vol. 18, No. 3, pp. 201–210, 2001.
3. B.L. Curless, New methods for surface reconstruction from surface range images, PhD dissertation, Stanford University, 1997.
4. K.H. Lee, S.B. Son, and H.P. Park, Generation of inspection plans for an integrated measurement system, *The 5th Biennial Conference on Engineering Systems Design & Analysis*, Montreux, Switzerland, July 10–13, 2000.
5. http://www.brownandsharpe.com.
6. K.H. Lee, H.P. Park, and S.B.A. Son, A framework for laser scan planning of freeform surfaces, *International Journal of Advanced Manufacturing Technology*, Vol. 17, No. 3, pp. 171–180, 2001.
7. http://www.jaewoo.com.
8. http://www.aracor.com.
9. http://www.siemensmedical.com.
10. H. Hoppe, T. De Rose, T. Duchamp, J. McDonald, and W. Stuetzle, Surface reconstruction from unorganized points, *SIGGRAPH 92 Proceedings*, University of Washington, Seattle, WA, pp. 71–78, 1992.
11. C.T. Loop, Generalized B-spline surfaces of arbitrary topological type, PhD dissertation, University of Washington, 1992.
12. K.H. Lee, H. Woo, and T. Suk, Point data reduction using 3D grids, *International Journal of Advanced Manufacturing Technology*, Vol. 18, pp. 201–210, 2001.
13. http://www.capture3d.com/.
14. http://www.pixar.com/.
15. http://www.geomagic.com.

8

Common Applications of 3D Printers

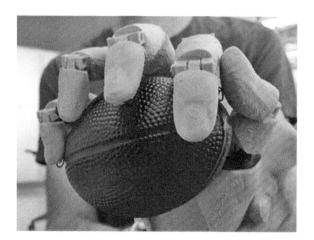

8.1 Introduction

3D printing (3DP) is a newly emerging field that is dependent upon continued research and development. These developments often include improvements to accuracy, build speed, and material properties. Projects in these fields have led to 3DP being used in various common applications. This chapter presents examples of projects by the author and his students at Loyola Marymount University in which the common applications of 3DP are explored. These projects include the application of 3DP with a bottle opener, a flower vase, the modeling of human faces, and the recreation of human fingers. These types of projects are useful to both the industry of 3DP and very instructive for those involved. The step-by-step procedures detailed in these projects also help to detail how a basic 3D software model can be converted to a functional physical object effectively. The further development in these applications leads to additional ways of how the 3DP could be utilized in everyday common design.

8.2 3D Modeling Software

In order to utilize the 3D printers effectively, models must be made in a 3D modeling software program that are eventually sent to the printer to be printed as physical objects. There are different types of modeling software programs used in the 3DP industry. These programs range from basic programs such as Tinkercad or Sculptris to more advanced programs like SolidWorks or Google SketchUp to the most robust design programs like Catia or Alias Design. A visual of these different levels of complexity can be seen in Table 8.1.

These different programs fall under the greater umbrella of computer-aided design or CAD. CAD is used in numerous different industries and is especially important for the 3DP industry. For the purposes of this section, all the projects detailed used the SolidWorks 3D modeling software program. SolidWorks is a common design software throughout the engineering and 3DP industry. The main interface of the SolidWorks software can be seen in Figure 8.1.

In this interface, the user can create a 2D shape that is dimensioned (such as the rectangle in Figure 8.1) and can be extruded to create a 3D model. An example of this extrusion process can be seen in Figure 8.2.

After the process of extrusion, additional elements can be sketched onto the different surfaces of the newly created 3D model and then extruded or cut from the initial model. As an example, a circle was drawn onto a side of the extruded rectangular prism and then cut through the object. The final version of these changes can be seen in Figure 8.3 as a finished part.

After finishing the design, extrusion, and cuts, the part can be prepared for 3DP. This is done by first ensuring that the object has closed sides and edges, which ensures that 3DP is possible. In addition, the user should check that the part does not have "zero-thickness" areas, where only a plane is present between two other features. SolidWorks does this automatically, but a final check is necessary to avoid errors in 3DP. Once the object is determined to be complete, the final step is to save the file as a stereolithography (STL) file. An STL file converts a model part into a series of triangles in order to 3D print. Triangles are an easy shape for 3D printers to follow; thus, the STL file format is best when preparing a model for 3DP. This is simply done by going to the "save as" tab and then clicking on the dropdown option under "file type" and selecting "*STL." Once this is done, the model is ready to be sent to a 3D printer and be printed as a physical object.

TABLE 8.1

Different Levels of 3D Modeling Software

Basic Software	Intermediate Software	Expert Software
Tinkercad	SolidWorks	Catia
Sculptris	Google SketchUp	Alias Design

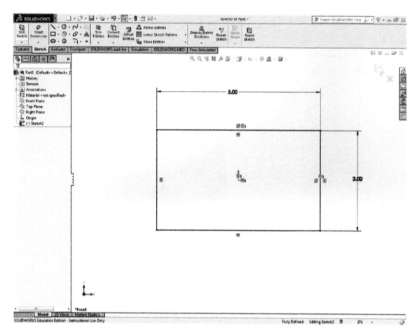

FIGURE 8.1
SolidWorks interface with basic shape drawn and dimensioned.

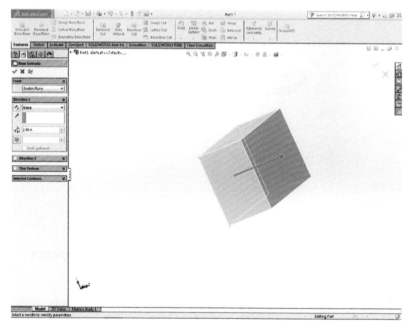

FIGURE 8.2
Extrusion of a basic shape in SolidWorks.

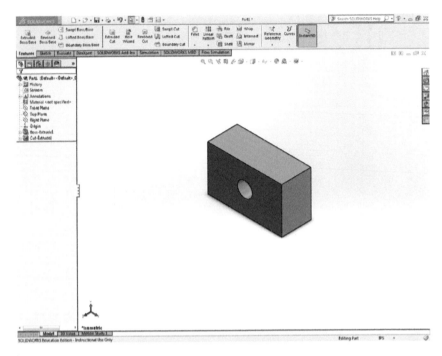

FIGURE 8.3
Basic part designed in SolidWorks.

FIGURE 8.4
CAD model of bottle opener.

8.3 Design and 3D Printing of an Everyday Bottle Opener

8.3.1 Introduction

A bottle opener is a practical everyday object that can be simply designed and 3D printed. In this project, a 3D CAD model was created using the SolidWorks software program (see Figure 8.4), which was then converted into an STL file, which was then sent to the MakerBot 3D printer and printed.

The material used to make the bottle opener was polylactate acid (PLA) plastic filament, which is heated to its melting point then pushed through a nozzle onto the build area. This process was repeated layer by layer until the bottle opener was finished. After this, the support material for the finished bottle opener was removed and then the bottle opener was ready to use.

8.3.2 Detailed Design

The process for this project had three main parts: the CAD modeling process, conversion of the CAD to an STL file, and then the actual printing of the bottle opener. The CAD modeling process began with finding the images of bottle openers and using them as a template. By using the template and researching the common dimensions of bottle openers, the CAD model was created in order to reflect the appearance and shape of an everyday bottle opener. In Figure 8.5, the dimensions of the finished model can be observed in the form of SolidWorks Drawing format.

After the CAD model of the bottle opener was finished, it was converted into an STL file.

Once the CAD model was converted to the STL file, the file was sent to the 3D printer. The MakerBot 3D printer was prepped before printing started. The preparation included leveling the build platform to ensure that the part had flat sides and ensuring that the platform was clean to avoid flaws in the final finished product. After this, the MakerBot's extruding nozzle was heated to 230°C, which is the best temperature for PLA plastic filament. After this, the MakerBot was ready for printing, which was achieved by simply

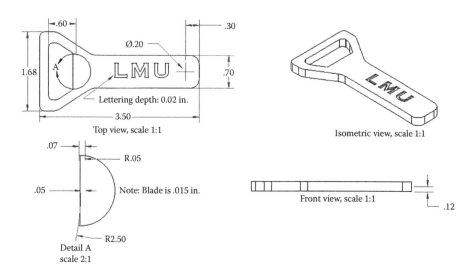

FIGURE 8.5
Drawing with dimensions of bottle opener.

pressing "print." The estimated time of print was 4h and 1min. The printer starting the print can be seen in Figure 8.6.

After printing, the bottle opener was removed from the print platform with tools like a putty knife and the excess support material was removed. The first version of the bottle opener was weak along the small blade and was quite thin for use. As a result, the small blade broke after testing the functionality of the bottle opener. The cracked blade of the bottle opener can be seen in Figure 8.7.

Following this flawed initial product, the bottle opener was redesigned. The blade was moved up to the top edge of the bottle opener, as well as its thickness being increased from 0.015 to 0.06in. In addition, the overall

FIGURE 8.6
MakerBot 3D printer beginning bottle opener print.

FIGURE 8.7
Bottle opener with cracked blade.

thickness of the bottle opener was changed from 0.1 to 0.2 in. These changes can be seen in Figure 8.8.

Once the changes were made to the bottle opener, it was printed a second time. The time for printing was 3 h and 25 min. Following this, the bottle opener was again removed from the print platform and the excess support material was removed. To improve the appearance of the bottle opener, sand paper was used to smooth the edges and remove any remaining excess material. After this, the bottle opener successfully opened a bottle and was ready for use. An image of the final version of the bottle opener can be seen in Figure 8.9.

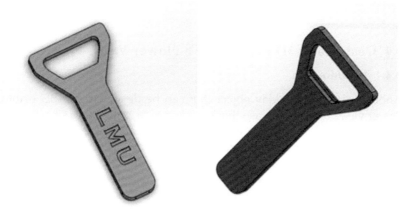

FIGURE 8.8
Edited bottle opener with moved blade and increased thickness.

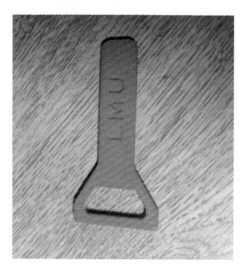

FIGURE 8.9
Final bottle opener.

8.3.3 Results and Conclusions

This project was meant to show how 3DP can be used to create an everyday object like a bottle opener. A small part was to be designed, converted to STL format, 3D printed with flaws first, redesigned to fix the errors, and finally 3D printed a second time successfully. Although the 3D printed bottle opener is not quite up to the standards and strength of its metal counterparts, it is still capable of opening a bottle. This project shows that with the basic knowledge of CAD, one can create objects found in everyday life.

8.4 Design and 3D Printing of a Flower Vase

8.4.1 Introduction

Another simple everyday object that can be designed and 3D printed is a flower vase. The flower vase that was designed is meant to sit around a glass of water or water bottle, in which the flowers are placed inside. The flower vase encasing the glass was created with geometric designs to make a more esthetically pleasing appearance. Although slightly more complex than a bottle opener, the process of the project was the same that includes 3D design using SolidWorks, conversion to an STL file, and finally printing.

8.4.2 Detailed Design

The first stage of this project was the 3D design of the flower vase in SolidWorks. This process began with creating a basic curve and then rotating the curve with a small thickness of 0.1 in. to create the original flower vase body. This body and curve were dimensioned to have a maximum diameter of 5 in. and a height of 9 in. Following the creation of the basic body, geometric designs and patterns were drawn on the body and cut out of the vase in order to create a more pleasing esthetic. The final product of the 3D design can be seen in Figure 8.10, as well as the dimensions of the flower vase in Figure 8.11.

After finishing the design of the flower vase, the file was converted to an STL file for printing. This STL file was much more complex than the bottle opener, as shown by the number of triangles used to model the object. The bottle opener used 5030 triangles compared to the 20,004 triangles used for the flower vase. This indicates that the flower vase is a much more complex object. Following this conversion to STL format, the file was uploaded into the MakerBot software. Within the software, it became apparent that due to time constraints and material, the vase could not be printed at full size. As a result, the vase was scaled down to 60% of its normal size. Once this was done, the vase was sent to the MakerBot printer to print. The time of

FIGURE 8.10
Design of flower vase in SolidWorks.

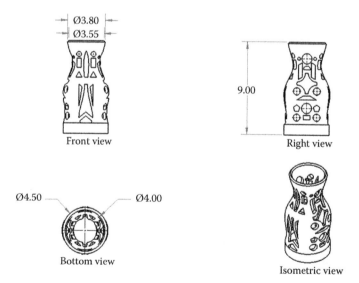

FIGURE 8.11
Drawing with dimensions of flower vase.

printing was 8 h and 15 min, which was more than double the print time of 4 h and 1 min for the bottle opener. Once the print was finished, the vase was removed from the MakerBot platform and sanded with sand paper to improve the appearance. After this, the flower vase was ready for use. The appearance of the final product can be seen in Figure 8.12.

FIGURE 8.12
Finished flower vase design printed at 60% size.

8.4.3 Results and Conclusions

The purpose of this project was to show how 3DP and design can be used to create slightly complex everyday objects such as a flower vase. Following the general process of design, conversion to STL format, and then printing, objects like flower vases can be easily designed and printed. Although the flower vase was printed at 60% of the normal size and does not function alone without a glass of water inside it, it is still capable of creating a more esthetically pleasing image than just the glass of water alone. From this standpoint, this project was successful in creating a functional everyday object for use.

8.5 Recreation of Human Face Using Reverse Engineering

8.5.1 Introduction

The goal of this project was to study the reverse engineering (RE) of complex objects such as a human's face using low-cost RE systems. The proposed work had three objectives: (1) recreating a physical model of the human face using a NextEngine 3D Scanner, (2) investigating the dimensional accuracy at which models can be produced by the system, and (3) evaluating the surface finish of the physical models. The goal for the evaluation was to create a model that was dimensionally accurate (<5% error).

Rapid prototyping (RP) is the fabrication of physical models of 3D computer-aided designs (CAD) through additive layer manufacturing techniques [1]. RP started in the late 1980s to produce models, prototypes, and functional end products in a shorter, more cost-effective way than the traditional subtractive methods such as milling or lathing. Recently, RE technology has emerged as a valuable tool in making parts for which there is no design data available, especially in the biomedical field. Examples of objects which have no design data include handmade prototypes, craftworks, reproduction of existing engineering objects from antiquity, and bones or teeth from the respective medical and dental industries [2,3].

RE is the process of duplicating an existing physical model and obtaining a 3D point cloud data file, which can be converted and printed using a 3D printer [4]. This is the technology by which companies have been able to reduce the cycle of product development and make design changes earlier in the process in order to reduce the cost [5]. It is known that noncontact, laser-based RE can reproduce the complex contours of the human face, but most of these systems are very expensive ($30,000–$50,000) [6]. This project concentrates on the evaluation of the effectiveness of a low-cost (<$5000) experimental setup: NextEngine's Desktop 3D Scanner and MakerBot Replicator 2.

The rapid acceleration of new and emerging technologies in RE is fueling a business growth in all aspects of design, product development, and manufacturing. Companies engaged in product development and manufacturing are in tremendous competition to bring a product to market faster, cheaper, with both higher quality and functionality. This competition has drastically decreased the price point of 3D printers and scanners. High-quality printers and scanners are now in the price range of most businesses, artisans, and consumers.

Little research has been conducted involving the recreation of the human face using low-cost RE systems due to the complex nature of facial features. For this reason, this project will focus on evaluating the effectiveness of utilizing a breakthrough low-cost RE systems to reconstruct the complex geometry of the human face. It is suggested that this project be extended to other science and engineering applications to serve as an important step toward RE objects of higher degrees of complexity.

8.5.2 Detailed Design

The experimental setup used for this project consisted of NextEngine's 3D Scanner, and MakerBot's Replicator 2 3D printer. Figure 8.13 shows the experimental setup of the preliminary project of a less complex football, and Figure 8.14 shows the experimental setup for the human face.

In the first part of the project, less complex objects were reverse engineered in order to determine the effectiveness of the RE systems using the NextEngine Desktop 3D Scanner. The initial objects of interest were a small

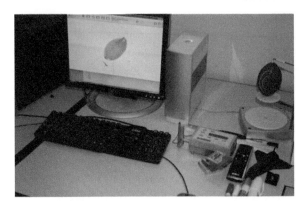

FIGURE 8.13
Experimental setup initial scan.

FIGURE 8.14
Experimental setup for face scan.

football. Multiple scans of these objects were conducted in order to construct 3D models. These 3D models were then printed using the MakerBot Replicator 2 3D printer as seen in Figure 8.15.

After these initial successful scans and 3D prints, the project moved toward its main goal of the recreation of the human face.

The same techniques were used in order to produce 3D models of a human face. From these scans, CAD files of a human face were created. An STL file and Point Cloud file were cleaned up and prototyped using MakerBot's Replicator 2 RP systems in order to fabricate three physical models of the

FIGURE 8.15
Initial object (Left) and 3D scanned and printed model (Right).

human face. The dimensions were measured to determine the margin of error between the original and reverse engineered objects. These measurements were used to rate the effectiveness of a low-cost RE system to recreate the complexities of the human face. It was clear from these measurements that the third 3D model produced had the greatest dimensional accuracy when compared to the initial model. This can be attributed to the technical challenges that were experienced throughout the preliminary stages of this project. The technical challenges faced during the execution of this project will be discussed further in results and discussion.

8.5.3 Results and Discussion

The face was scanned several times in different orientations. These different orientations captured images of various locations on the face as seen in Figure 8.16. The gaps in some scans were filled in by scans in other orientations

FIGURE 8.16
Preliminary facial scans.

and overlaid creating a single model of the entire face. The scanned face was successfully enhanced using computer software and converted into both a point cloud file and STL compatible for 3DP as seen in Figures 8.17 and 8.18.

The computer model of the human face showed minor flaws when compared to the original as well as uneven transitions where the different scans were stitched together using the computer software.

FIGURE 8.17
Point Cloud file of preliminary scan.

FIGURE 8.18
STL file of preliminary human face scan.

An initial physical model was constructed, but the model produced was not recognizable as a human face. This failure was attributed to the inability of the NextEngine software to, without any manipulation, create a viable STL file. The scans were then significantly worked on using the NextEngine software until an STL file was created that resembled the contours of a human face. This model still contained several minor flaws, which were attributed to the technical challenges experienced during this stage of the project. These included unexplainable gaps and holes in the printed 3D model, as well as areas for which additional material was laid. These problems were caused by the inability of the NextEngine software to take the scanned images and properly stitch them together in a way that the 3D printer could easily interpret. In this case, the 3D printer could not correctly create an STL file, which resulted in the absence and addition of material to the physical model. The third model required additional scans to fill in and smooth out the point cloud file.

After significant manipulation using various tools in the NextEngine software many of the problems experienced in the second iteration were rectified in the final 3D printed model. The final physical model produced using the MakerBot Replicator 2 showed drastic improvements in surface transitions and overall quality. This was attributed to overcoming the technical challenges experienced in the second model by using a greater number of scans to create the STL and Point Cloud files as well as refining the surface of the face using the NextEngine scanning software. The measurements of the dimensions of the original and prototyped objects are summarized in Table 8.2.

TABLE 8.2

Human Face Dimensions (0.75 Scale)

Nominal (in.)	Measured (Mean, in.)	Percent Error
Length Dimensions, MakerBot Replicator 2		
7.5	5.5	26.6%
Width Dimensions, MakerBot Replicator 2		
4.25	3	29.4%
Nose Height Dimensions, MakerBot Replicator 2		
1.25	1	20%
Nose Width Dimensions, MakerBot Replicator 2		
1.5	1	33.33%
Right Eye Dimensions, MakerBot Replicator 2		
1	0.75	25%
Left Eye Dimensions, MakerBot Replicator 2		
1	0.75	25%
Lip Dimensions, MakerBot Replicator 2		
2	1.25	37.5%

This table shows the prototyped dimensions according to length, width, nose height, nose width, right eye, left eye, and lips. The percent errors of these dimensions are also displayed for the human face. As the table illustrates, the first prototyped model was very accurate when capturing particular features and much less accurate when capturing others. More accurate features (below 7%) were the length, width, nose height, and eyelid dimensions of the face. The least accurate features by far were the lip dimensions and nose width dimensions, which had percent errors upward of 16.67%, and 33.33%, respectively. On the other hand, the second prototyped model possessed extremely accurate features across all parts of the face, the highest error reaching a mere 4%. The dimensional accuracy of the second human face model is displayed in Table 8.3.

Although most dimensions of the prototyped model were very accurate, the surface finish of the second model was much worse than that of the third and final model. The surface of the second prototyped face contained areas of smooth finish, while others were much rougher. The forehead of this prototype was the most consistent and smooth surface of the model. In contrast, the areas around the eyelids and lips contained the most irregular surface finish of the model. In addition, extra material was laid on the inner section of the right eyelid that was not shown in the CAD models of the human face prior to 3DP. Although the initial models showed these flaws, the third 3D model had a uniformly smooth surface across the entire face and did not contain any unexplained holes or added material in its surface.

TABLE 8.3

Human Face Dimensions (Full Scale)

Nominal (in.)	Measured (Mean, in.)	Percent Error
Length Dimensions, MakerBot Replicator 2		
7.5	7.5	0%
Width Dimensions, MakerBot Replicator 2		
4.25	4.25	0%
Nose Height Dimensions, MakerBot Replicator 2		
1.25	1.20	20%
Nose Width Dimensions, MakerBot Replicator 2		
1.5	1.5	0%
Right Eye Dimensions, MakerBot Replicator 2		
1	1	0%
Left Eye Dimensions, MakerBot Replicator 2		
1	1	0%
Lip Dimensions, MakerBot Replicator 2		
2	2	0%

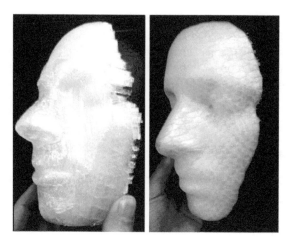

FIGURE 8.19
Second and third physical model of human face.

These errors are thought to be not due to enough point cloud data around the areas of complex geometry changes. In order to create a better model, more scans were taken from new angles to create a 3D model with fewer errors. Two more scans with higher degrees of angle were added to the third that were previously used in order to rectify the surface finish errors. This is demonstrated by the increased dimensional accuracy in the final 3D model of the human face. The improvement in the dimensional accuracy of the facial models can be seen in Figure 8.19.

While the NextEngine scanner captured many of the main facial features very accurately in the final model, it is believed that a more expensive scanner designed specifically for capturing human features is necessary for getting finer details such as eyebrows, lip definition, and hair.

8.5.4 Conclusion

The following conclusions were drawn from the project thus far:

- A low-cost system is able to reverse engineer complex objects with only minor limitations.
- Significant time and effort is still required to create the complex contours of an object.
- The primary limitation of the NextEngine Desktop 3D Scanner is retrieving data from areas of the human face containing many features (contours of nose, eye sockets, lips).
- Preliminary results show minor flaws in the reverse engineered model where scans were stitched together using the NextEngine software.

Future developments in this field ought to be pursued using alternative scanners and 3D printers in order to compare speed, efficiency, and additional constraints. This project should be extended to other science and engineering applications in order to be able to reverse engineer objects with great degrees of complexity.

8.6 Recreation of Human Fingers for Accident Victims

8.6.1 Introduction

The goal of this project was to study the ability of low-cost RE and 3DP systems to recreate a complex object such as a human hand so that it could be used for biomedical applications. The proposed work had three objectives: (1) the first objective was to recreate the physical model of the human hand using a low-cost experimental setup (<$5000), (2) the second objective was to assess the ability of the reverse engineered hand to perform common tasks of everyday life, and (3) the third objective was to investigate the potential biomedical applications of the reverse engineered human hand.

3DP, also referred to as RP and additive manufacturing, is a manufacturing method in which 3D objects are made by fusing or depositing material in layers. Today, materials range from plastics, metals, ceramics, and powders, to even liquids and in some cases living cells [7]. This method of building objects layer by layer started in the late 1980s as a way to produce models, prototypes, and functional end products quicker and cheaper than ever before. Recently, RE has been partnered with 3DP techniques to become a valuable tool in making parts for which there is no design data available, especially in the biomedical field. For the purposes of this project, examples of objects for which no design data exist include human limbs, organs, bones, and even living tissue.

There are many individuals who face living without a limb. An individual may have lost his or her hand or fingers in a freak accident, in war, or due to a disease. Others include children who have been born without a hand or possess an underdeveloped hand that they cannot properly use. For any of these cases, the creation of a functional human hand for them to use can be very difficult and highly expensive. This is especially true for children who will have to replace their prosthetic quite often during years of growth. Recent news has shown a spotlight on the ability of 3DP to create prosthetic limbs quickly, cheaply, and effectively. While many prosthetics can cost as much as $30,000 [8], a low-cost reverse engineered limb is only on the magnitude of a few hundred. Although others have created inexpensive prosthetics using 3DP technology, little research has been done using low-cost RE to create both inexpensive and life-like limbs. For this reason, this project focused on evaluating the effectiveness of using low-cost RE systems as partners with existing 3DP devices to reconstruct the complex geometry of the human hand. The model was evaluated by its ability to perform common everyday tasks such as picking up a cup/bottle, holding a pen/pencil, or opening/closing around an object. It is suggested that this project be extended to other biomedical areas in the future to include the RE of other human limbs such as toes, feet, arms, and legs as well as human organs. This project hopes to serve as an important step toward exploring the possibilities of producing biomedical devices using low-cost RE and 3DP techniques.

8.6.2 Detailed Design

The experimental setup used for this project consisted of a Life Casting Starter Kit, a NextEngine 3D Laser Scanner, and a MakerBot Replicator 2 Desktop 3D Printer. Figure 8.20 shows the experimental setup for recreating the casted hand into a 3D digital file.

The first portion of this project focused on determining the best method in which to scan the physical hand trying to be recreated. There were two methods investigated: (1) scan a live subject's hand in real time and (2) create a casted version of the subject's hand that could be placed on a scanning platform. Due to the experimental setup and time given for this project, the second method was determined to be the most effective at producing a high-quality reverse engineered hand. A kit designed for casting human limbs was used to mold and cast the chosen test subject's hand. This process was repeated several times using different molding kits (Precious Impression Molding Kit and Life Casting Starter Kit) available on the market in order to ensure a high-quality casting was achieved. The casting kits used for this research averaged approximately $30 dollars in price. Once a good quality cast of the subject's hand was obtained, it was scanned in multiple orientations in order to construct a digital 3D model using NextEngine's Scan Studio Software. This software was used to stitch each of the individual scans together to produce a water tight CAD model with no holes and minimal imperfections. After the scans of the hand were successfully stitched together, the next step of this project was to convert and construct the 3D printed model. Figure 8.21 shows the 3D Scan Studio data file of the reverse engineered hand.

In order to construct a fully functioning printed hand that could be articulated and used to perform everyday functions, the digital 3D model created in Scan Studio was converted from a Scan Data File to an Initial Graphics Exchange Specifications (IGES) data file. The IGES file format is what allowed the 3D computer model to be imported into SolidWorks and altered to mimic the joints of human hands. This required segmenting each finger into three parts (excluding the thumb). 45° angles were cut where the joints would bend and mate with each other to leave room for their natural motion. Additionally,

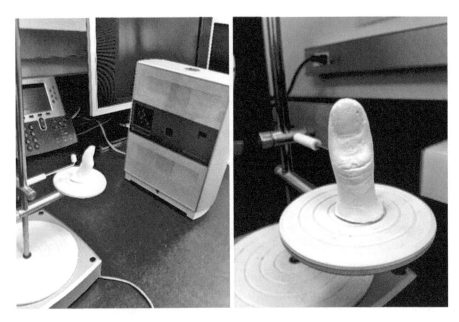

FIGURE 8.20
Experimental setup for laser scanning.

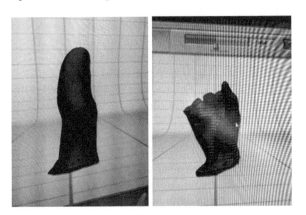

FIGURE 8.21
3D models in Scan Studio.

splines were drawn to cut custom pathways through each section of the fingers in order to allow for a cable to pass through them. This was the foundation for the cable-driven system used to open and close the recreated hand. Finally, from the redesigned CAD file, the reverse engineered human hand was 3D printed using a MakerBot Replicator 2. The hand was ultimately printed as a single assembly which did not require any separate pieces to be assembled together to produce the recreated human hand. Figure 8.22 shows one variation of the articulating joint design tested for its versatility and durability.

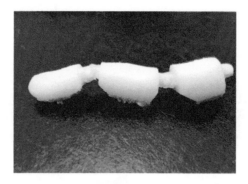

FIGURE 8.22
Single articulating finger assembly.

The effectiveness of the low-cost experimental setup to reproduce a functioning human hand was evaluated by assessing its ability to perform everyday tasks. The specific tasks chosen for the purpose of this project were: (1) the ability to pick up a cup or water bottle, (2) the ability to pick up/hold a pen or pencil, and (3) the ability to easily close and open the hand around an object (door handle/ handle bars/steering wheel/toy). The technical challenges and final results of this project are presented in the results and discussion section of this paper. Additionally, potential applications of reverse engineered human limbs are explored in further detail.

8.6.3 Results and Discussion

The casted human hand was scanned in many orientations in order to capture the most complete 3D image of the hand as much as possible. Each individual scan was used to capture a different side/location of the hand to complete the digital model. After the scans were stitched together using Scan Studio Software, the remaining gaps and holes were filled in using the "fill" application and "water tight model" setting in the program. The default file format for the Scan Studio software is in a scan data file format. This file format was converted to an IGES data file format so that it could be imported into the CAD software, SolidWorks. Manipulating the digital model in SolidWorks allowed joints to be constructed and the foundations of the cable-driven motion system to be laid (See Figure 8.23).

The initial casting of the hand created using a commercial casting kit by Precious Impressions produced high-quality fingers, but unfortunately did not create a complete cast of the subject's hand. This was a result of the kit's primary function as a children's hand molding kit, not intended for adult-sized limbs. Fortunately, the quality of the individual finger castings allowed them to be useful in obtaining the required scans for each of the fingers and were ultimately used in the final digital hand model and assembly. An additional casting kit by Life Casting was purchased and used to create the cast

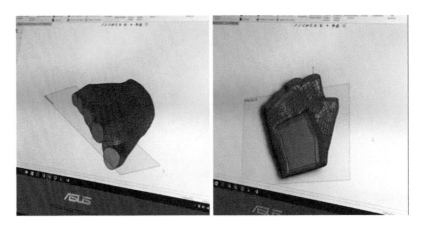

FIGURE 8.23
CAD model in SolidWorks.

of the rest of the subject's hand up to where the hand and arm meet at the wrist. This kit featured Alja-safe molding gel, which inherently helped to produce a better quality mold. A separate set of scans were taken of just the main body of the hand from the second cast and were stitched together to create a digital model. Due to the quality of the castings used for this project, only very minimal flaws were present in the digital model and were easily cleaned up using the Scan Studio Software.

After the digital model of the human hand was created and refined in Scan Studio it was converted from a scan data file to an IGES data file that was compatible with SolidWorks CAD software. Initial attempts were made to convert the scan file to an STL file format, but this merely created an image inside SolidWorks that could not be changed or manipulated. After seeking advice from NextEngine and online 3DP forums, the best file format for this project was determined to be the IGES data file. Each finger and the body of the hand were then imported into SolidWorks as separate "parts" that could be handled as separate entities of the entire assembly. Once the joints for each finger were constructed and the pathway for the cable system was integrated into the design, the fingers alone were printed for testing and redesign.

While each finger posed its own challenges due to size variations and orientation, the same basic approach was taken to make each finger a fully functioning limb that could later be implemented into the overall hand's construction. To create the functioning joints in SolidWorks, many lines and planes needed to be created. Due to the nature of the scan data, a sketch was not able to be created on the surface of a finger. To get around this problem, precisely placed planes were created to be able to make cuts on the finger and separate the fingers into separate parts. These cuts made it possible to create sketches directly on the fingers. Once the sketches were made, an extrusion of part of the joint could be produced. After all of the joints are created, an assembly was generated to join all of the parts of the fingers together. The primary technical challenge is to overcome the design of the individual fingers was the implementation of robust, yet free moving joints that could not only be printed as single assemblies but also withstand the stresses of everyday activities. This technical challenge was overcome through an iterative design process in which a single finger was designed, printed, tested, and redesigned until an acceptable design was found. This included testing of a snap-in, press-fit, and bracket and peg joint design. Once the style of joint was determined, pathways for the cable system were created through each finger to facilitate the opening and closing motion of the prosthetic. Finally, the fingers were integrated into the body of the hand, which required much less rework and manipulation, and the entire hand assembly was printed on Loyola Marymount's MakerBot Replicator 2 3D Printer.

Measurements were taken of the subject's hand, the casted model, and the 3D printed prosthetic in order to determine the accuracy at which the low-cost system could recreate human limbs, the ultimate goal being to recreate a prosthetic that is as lifelike as possible and lend itself to the principle of mass customization for each user. The dimensions were recorded and compared in Tables 8.4 through 8.6. These tables show a comparison in the overall width of the wrist and length of each finger. The percent errors are also shown for each dimension.

Tables 8.4 and 8.5 show how the dimensions of the subject's hand were translated throughout the casting, RE, and printing process. Specifically, Table 8.4 compares the dimensions of the casted hand to the subject's hand. Although several of the measured dimensions showed low percent errors (<7%), a few showed more errors (>10%). This can be attributed to the position in which the mold had to be casted using the kit that was purchased. In order to fit the subject's hand in the casting container, the subject was forced to slightly bend his or her fingers. This significantly influenced the ability to accurately measure the "straight" length of each finger, leading to the larger percent errors. Similarly, Table 8.5 compares the dimensions of the printed prosthetic to the casted model. Again, several measured dimensions had low percent error (<6%), while others were much greater (>14%). It can be seen that the dimensional errors occurred on the same three parts of the models, the pointing finger, the middle finger, and the ring finger. This is once again attributed to the slight bend that had to occur in the fingers in order to cast the hand in the kit's container.

As Table 8.6 clearly shows, the printed model was very accurate (<10% error) when recreating the life-like dimensions of the subject's hand. This lends itself to the possibility of low-cost mass customization where one could have their limb recreated to nearly exact dimensions. This is unlike other low-cost printed prosthetics on the market which are bulky and angular. For this reason, a primary goal of this project was to create a low-cost prosthetic that replicated not only the motion of a human hand but also its organic shape, so as to be as life-like as possible.

In addition to being dimensionally accurate, the hand was evaluated on its ability to perform everyday tasks such as picking up a pen or pencil, picking

TABLE 8.4

Casted Hand Dimensions vs. Nominal Hand Dimensions

Measured (mm)	Nominal (mm)	Percent Error (%)
Wrist Dimensions		
55.82	53.22	4.88
Thumb Dimensions		
57.87	56.77	1.94
Pointer Finger Dimensions		
63.86	73.29	−12.86
Middle Finger Dimension		
69.48	79.99	−13.14
Ring Finger		
63.90	72.34	−11.67
Pinky Finger Dimensions		
59.57	55.76	6.85

TABLE 8.5

Printed Hand Dimensions vs. Casted Hand Dimensions

Measured Printed (mm)	Measured Casted (mm)	Percent Error (%)
Wrist Dimensions		
52.98	55.82	−5.08
Thumb Dimensions		
60.21	57.87	4.04
Pointer Finger Dimensions		
80.21	65.86	21.79
Middle Finger Dimensions		
79.75	69.48	14.79
Ring Finger Dimensions		
78.83	63.90	23.35
Pinky Finger Dimensions		
60.85	59.57	2.14

up a cup or bottle, and gripping objects such as toys, door knobs, handle bars, and steering wheels. The first object tested was a miniature football that a child using this prosthetic may want to pick up to play with his or her friends. When tensioned, the prosthetic was able to reach down and pick the toy up off the ground or off a table. Only minor challenges were experienced

TABLE 8.6

Printed Hand Dimensions vs. Nominal Hand Dimensions

Measured (mm)	Nominal (mm)	Percent Error (%)
Wrist Dimensions		
52.98	53.22	−0.44
Thumb Dimensions		
60.21	56.77	6.06
Pointer Finger Dimensions		
80.21	73.29	9.45
Middle Finger Dimensions		
79.75	79.99	−0.29
Ring Finger		
78.83	72.34	8.96
Pinky Finger Dimensions		
60.85	55.76	9.13

when picking up an object of this type due to its shape, size, and high-grip material. This is shown in Figure 8.24.

The second object used for evaluating the effectiveness of the printed prosthetic limb was handle bars. This type of object would be very important for a child who wants to ride his or her bike, as well as adults who require gripping the steering wheel of their car while driving. Once again, when tensioned, the prosthetic was able to completely grab the handle bar and could even push and pull the handle bar to simulate steering a bicycle. Similar to the toy football, the hand was able to function with only minor challenges due to the size, shape, and material properties of the handle bar. This is displayed in Figure 8.25.

FIGURE 8.24
Prosthetic hand picking up toy football.

Unfortunately, when it came to trying to grab or pick up smaller objects such as pens and pencils or plastic objects such as water bottles and cups, the prosthetic hand was unable to close around the objects enough to successfully pick them up and use them. This is due to technical challenges experienced with the cable motion system, the articulation of the finger joints, and the plastic material used for the prosthetic itself. First, it was observed that the cable system would need to be redesigned in the future to improve the hand's ability to fully close around smaller objects. Second, it was observed that the joints tended to catch the hand that was closed tighter and therefore they should be improved in future iterations to better replicate the organic motion of a human hand. Finally, it was observed that the PLA plastic used to print the prosthetic with the MakerBot 3D printer presents problems when trying to grip smaller objects or smooth plastic objects due to the lack of friction to keep them in place. Therefore, other materials should be investigated to prevent objects from slipping out of the prosthetic when in use.

The bioengineering field has grown drastically recently due to the enhancement of crucial technology. This technology includes 3D scanners, 3D printers, and new revolutionary materials. 3D scanners have made it possible to reverse engineer extremely complex anatomy. This not only helps with prosthetics, but also with any type of doctor that must perform a complex procedure on a patient. With the use of the 3D scan, 3D printers are able to create a working model, which would be impossible on current types of manufacturing tools, like a CNC machine. As the 3D printer gets more popular, more materials are able to be printed with them. It is possible to print with a highly flexible polyurethane material, which would be useful for the future of prosthetics. For instance, a new company called Ninja Flex has come out with a type of material which can be used in the joints of prosthetics so that the highly elastic properties can help the user open his or her hand more

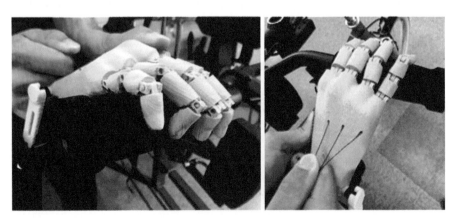

FIGURE 8.25
Prosthetic hand gripping bicycle handle bars.

easily after closing it. It is easy to see that the future of prosthetics will blend this technology with robotics to create more precise moving and better looking limbs. Another example is a new type of smart material that can be flexible under zero voltage and stiff when voltage is applied to it. This type of material would be perfect for making prosthetic limbs more lifelike. While the prosthetic hand produced by this project serves as a stepping stone for low-cost mass customized prosthetics, there is still much that can be done to enhance this project to extend to other limbs and biomedical applications.

8.6.4 Conclusion

The following conclusions were drawn from the project thus far:

- A low-cost system is able to reverse engineer a human's hand with minimal dimensional errors and minor limitations on motion.
- Significant time and effort is still required to perfect the motion system of the hand and how it will integrate with a comfortable lightweight mount for the user.
- The primary limitations of this low-cost setup are the materials that can be used for printing (i.e., PLA and acrylonitrile butadiene styrene plastics) as well as retrieving data through "live" RE as opposed to creating a casting of the subject's hand.
- Preliminary results show that this type of prosthetic is better for large objects made from materials with inherent friction due to their surface finish. Additional research and redesign are necessary for being able to pick up smaller and smoother objects at this time.

Future developments in this field should be pursued using alternative scanners, 3D printers, and base materials. These would improve the way in which an amputee or accident victim could have his or her hand recreated in real time as well as the effectiveness of the prosthetic to pick up a wider range of objects. It is suggested that this project be extended to the recreation of additional limbs such as arms, legs, and feet, especially as technology in this field rapidly improves to make recreation quicker and cheaper than ever before.

8.7 Summary

This chapter has attempted to show how 3DP and RP are commonly applied through various projects executed by students and faculty of Loyola Marymount University. From the basics of creating everyday objects like a bottle opener or a flower vase to more complex medical applications like facial reconstruction and the recreation of a human hand, there are various different fields that utilize 3DP. Most of these applications follow the same process of choosing a model, recreating the model in modeling software (SolidWorks), saving the file as an STL file, printing the object on a 3D printer, and finally postprocessing to improve appearance. Although there are small differences in each individual project, most common applications for 3DP follow this basic five-step procedure. Due to this relatively easy and basic process, 3DP is being used more and more everyday and has a bright future in being utilized in numerous different common applications.

8.8 Questions

1. Mention several software packages that you can use to design a part.
2. What is the difference between a conventional design software and a parametric design software?
3. What is a mechanical design? Describe the steps involved in the design process.
4. Design an everyday coffee cup. Show different views of the design (top view, side view, front view, and parametric view).
5. Describe the steps you take to prototype the cup.
6. What type of 3D printer you will use and why?
7. What is the role of 3DP in biomedical applications?

References

1. R. Noorani, *Rapid Prototyping-Principles and Applications*, John Wiley & Sons, Hoboken, NJ, 2006.
2. G. Lin and L. Chen, An intelligent surface reconstruction approach for rapid prototyping manufacturing, *The Fourth International Conference on Control, Automation, Robotics and Vision*, Singapore, 43–47, December 3–6, 1996.
3. C. Schoene and J. Hoffmann, Reverse engineering based on multi axis digitized data, *Proceedings of the International Conference on Manufacturing Automation*, Hong Kong, 909–914, April 28–30, 1997.
4. V. Raja, *Reverse Engineering an Industrial Perspective*, University of Warwick, Warwick, UK 2008.
5. S. Singh, Rapid reverse engineering to rapid prototyping: A case study, *Proceedings on Reverse Engineering, SME*, Newport Beach, CA, December 9–10, 1998.

6. C. Boehnen and P. Flynn, Accuracy of 3D scanning technologies in a face scanning scenario, University of Notre Dame, 2005.
7. C.L. Ventola, Medical applications for 3D printing: Current and projected uses, *Pharmacy and Therapeutics*, Vol. 39, pp. 704–11, 2014.
8. G. McGimpsey and T. Bradford, Limb prosthetics services and devices, Bioengineering Institute Center for Neuroprosthetics (n.d.): 11. Print. Reverse Engineering, SME, Newport Beach, CA, December 9–10, 1998.

9

3D Printing in Medicine

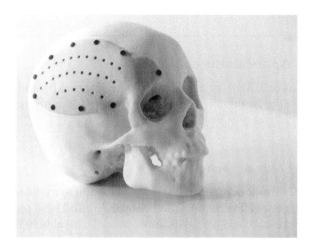

9.1 Introduction

3D Printing (3DP) is a process that is used for fabricating solid prototypes from a computer-aided design (CAD) data file. The field of 3DP has become one of the fastest growing fields in the engineering industry today. One area in 3DP that is progressing quickly and one which 3DP is having a pronounced impact is that of the medical field. Using 3DP technologies in the field of medicine is no simple task though. It is a complex process and requires a multidisciplinary approach with collaboration between doctors and engineers. The number of medical applications of 3DP is increasing everyday. The uses of 3DP can be classified in many categories including creation of customized prosthetics, implants, anatomical models, and tissue and organ fabrication. In addition, research is being done on drug dosage forms, delivery, and discovery. The benefits of uses of 3DP in medicine are many including the customization and personalization of medical products, drugs, equipment, cost-effectiveness, increased productivity, and the customization of design and fabrication. The benefits of 3DP in the medical field are numerous. It has the potential to help cut costs in medical procedures, save time during surgeries, and even help end the donor shortage.

This chapter describes the many uses of 3DP in medicine, types of medical imaging, software for medical models, medical materials, and a case study. The case study will describe a detailed procedure for printing c1 vertebrate from the CT scan data.

9.2 Medical Applications of 3DP

The applications of 3DP in medicine are expanding rapidly and are expected to revolutionize the healthcare industry. A number of applications in which 3DP has had an effect will be discussed next [1].

9.2.1 Teaching Aids and Simulators

The rapid prototyping (RP) may be used as a means to make models of any given body part or even recreate a medical condition. These models provide useful tools for researchers and educators. Models are able to be distributed in kits to schools, and provide a hands-on illustration of certain body parts, allowing the students to practice procedures or gain a better understanding of the human anatomy. Being able to look at a 3D model of a bone is a much better learning tool than staring at a picture of it in a book. These 3DP models may be used in medical schools and in training courses, allowing students to perfect their skill without causing harm to an actual person or using cadavers.

9.2.2 Presurgical Planning Models

Surgery tends to be a complex and time-consuming procedure. Through the use of 3DP, surgical procedures can become safer as well as less time-consuming. Models of a patient's anatomy may be created through 3DP methods and then used to rehearse complex procedures. Though doctors practice and study each surgical procedure they perform, no person is exactly the same, thus no surgery is exactly the same. Taking a CT scan of the patient allows the doctor performing the surgery to rehearse with that particular patient's exact anatomy, effectively reducing the actual time spent in surgery and preparing the doctor for surgery. Surgeries that were previously inoperable or extremely difficult and potentially dangerous have become a more accessible option for doctors. Just as an actor in a play, it may not always be a perfect performance the first time; it takes rehearsal to get every line and movement right. 3DP allows for doctors to go through their own rehearsal, ensuring each surgery is executed to perfection. The models may also be used by the doctors as visual aids in order to better plan a surgery, increasing the chances of a successful surgery. In a recent case, doctors find a tumor in a woman's brain and proposed an invasive surgery. The man's wife sought a less-invasive procedure, so he used magnetic resonance imaging (MRI) files to create models of her skull and then mailed them to the surgeon who eventually performed the surgery. By analyzing and practicing on the model of the woman's skull, the surgeon was able to remove the tumor through a small opening, a much less-invasive procedure. The benefit of using 3DP in this application is only exemplified in the case that a patient has any anatomical abnormalities or deformities.

9.2.3 Customized Surgical Implants

Through the use of 3DP, customized surgical implants may be created for a patient. Currently, implants are typically standardized, and a patient is left on the surgery table while the implant is being customized to fit. However, with 3DP scan of the patient may be done and then a customized model created that already matches the anatomy of the patient. This allows for the customized implant to be made beforehand, decreasing the amount of time spent in surgery and reducing the risk of surgical complications.

9.2.4 Mechanical Bone Replicas

Bones are not solid homogeneous, but have two distinct regions: the cortical and trabecular bones. 3DP allows for the creation of a nonhomogeneous model that accurately reflects these two regions, providing a replica that is more true to life. A lattice structure of an stereolithography apparatus (SLA) can be used to create these two distinct regions and provide mechanically correct bones. This allows doctors to observe the bone strength under various conditions and recreate events that may cause fractures, stresses, and other changes in the bone.

9.2.5 Prosthetics and Orthotics

The use of 3DP in prosthetics or orthotic devices is different from conventional methods because it starts with each individual patient's anatomy, making it one of the most obvious fields that 3DP can benefit. As in other applications, medical image data files may be used to study the patient's anatomy and create designs that already account for the patient's specific alignment characteristics. This reduces the number of times a prosthetic needs to be refitted, cutting down cost. The use of 3DP allows for a prosthetic or orthotic that has biomechanically correct geometry to be created the first time and provides the patient with an improved fit, comfort, and stability.

9.2.6 Bioprinting

Bioprinting is a term that has been referring to 3D printing actual living organs and tissues. In this process, the first step is to take a biopsy of the organ or tissue to be replaced. This biopsy is examined and certain cells with regenerative potential are isolated and multiplied. These cells are then mixed with a liquid material that provides oxygen and other nutrients in order to keep the cells alive. This mixture of cells is then placed into a printer cartridge, while a separate printer cartridge is filled with a biomaterial that will be printed into the organ- or tissue-shaped structure. The structure itself is created on the computer through the assistance of a patient's medical scans. Once the printer is ready to print, it builds the structure layer by layer while it embeds the cells into each layer. If the cell is provided with the right mixture of nutrients, growth factors, and placed in the right environment, it will grow and perform its functions. This function of RP to print organs is only a scaled-up version of current processes to grow organs in laboratories. However, using 3DP allows for the process to be more precise and reproducible. The ability to print organs and tissues is a huge breakthrough in the 3DP world and one that has many benefits, and the goal of this application is to help solve the shortage of donor organs. 3DP organs also have a huge benefit over donor organs in the sense that the organs are engineered with a patient's own cells, thus eliminating the risk of rejection. The ability to 3DP organs for a patient is one of the most exciting breakthroughs in recent years; it has the ability to end the shortage of organ donors and save many lives.

In addition to the aforementioned applications, 3DP is used to print organs, stem cells, skins, heart, and blood vessels. In the same way as the tissue and organ cells are printed and studied, cancer cells are also being bioprinted for effective studies.

9.3 Types of Medical Imaging

Modern medical imaging technology has made a tremendous impact on patient diagnostic procedures over the last twenty-five years. The 3DP technology that operates in a CAD environment provides many advantages for medical applications. The virtual images created by the medical imaging equipment can be converted to 3D models for the medical professional to visualize different potential scenarios before an operation is performed.

Medical imaging first started with X-rays as a tool for seeing through the flesh and bones of a patient. The availability of sophisticated computers and improved digital image processing has resulted in improved imaging technology. An innovative technology in medical imaging uses the traditional methods to create 3D models. These 3D models of the human body and its organs are then sliced, rotated, and analyzed depending on the area of interest. The 3D models are constructed by taking slices of the desired body part and then stacking them together to form the full-scale model. Figure 9.1 shows a simplified version of the technique. Both computers and conventional imaging methods are used to accomplish the image acquisition.

Classical imaging techniques are identified through the 2D images they produce. There are several types of conventional medical imaging techniques, such as X-rays, computed tomography (CT), MRI, and positron emission tomography. In this section, we shall briefly describe the operating principles of these techniques.

9.3.1 X-ray Technology

X-rays are used everyday worldwide by medical professionals to look at bones that may be broken or contain stress fractures, to see something inside the body that might have been swallowed, to see images of gunshot wounds, to view gallstones, or even to scan baggage at the airport.

X-rays are penetrating electromagnetic radiations, having a shorter wavelength than light, and are usually produced by bombarding a target with high-speed electrons. X-rays were discovered accidentally in 1895 by a German physicist, Wilhelm Conrad Roentgen [2] while he was studying cathode rays in a high-voltage gaseous discharge tube. Despite the fact that the tube was enclosed in a black cardboard box, Roentgen observed that a barium-platinocyanide screen, inadvertently lying nearby, emitted fluorescent light whenever the tube was in operation. Upon further investigation, he determined that the fluorescent light was caused by invisible radiation of a more penetrating nature than ultraviolet rays. He named the

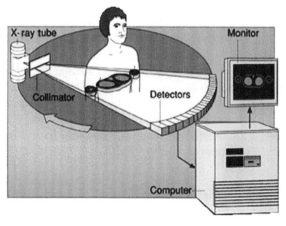

FIGURE 9.1
Simplified model for 3D image acquisition.

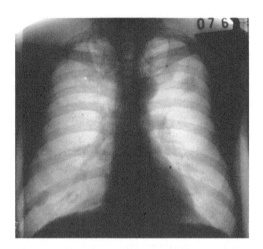

FIGURE 9.2
X-ray of a human chest cavity.

invisible radiation "X-ray" because of the unknown nature of the radiation. In Figure 9.2, an X-ray image of a human chest is shown.

9.3.2 Nature of X-rays

X-rays are electromagnetic radiations whose wavelengths range from $100\,\text{Å}$ to $0.01\,\text{Å}$ ($1\,\text{Å}$ is about $10^{-8}\,\text{cm}$). The shorter the wavelength of the X-ray, the greater is its energy and its penetrating power. X-rays are produced whenever high-voltage electrons strike a material object. While most of the energy of the electrons is lost in heat, the remaining energy produces X-rays by causing changes in the objects atoms as a result of the impact.

The X-rays affect a photographic emulsion in the same way as light does. Absorption of X-radiation by any substance depends on its density and atomic weight. The lower the atomic weight of the material, the more transparent it is to X-rays of a given wavelength. For example, when the human body is X-rayed, the bones consisting of higher atomic weights than the surrounding flesh absorb radiation more effectively and cast dark shadows on a photographic plate. Neutron radiation, which is different from X-radiation, is now also used in some types of neutron radiography producing the almost exact opposite results. Objects that cast dark shadows in an X-ray picture are almost always light in a neutron radiograph.

9.3.3 Applications of X-rays

Research has shown that X-ray technology has played a very vital role in theoretical physics, especially in quantum physics. Using X-rays as a research tool, scientists have been able to confirm experimentally the theories of crystallography. The X-ray diffraction method has been used to identify crystalline substances and to determine their structures. Virtually all modern crystalline substances were either discovered or verified by X-ray diffraction analysis. By using X-ray and diffraction methods, chemical compounds can be identified and the size of the ultramicroscopic particles can be established. Chemical elements and their isotopes are regularly identified by X-ray spectroscopy.

Some recent applications of X-rays involve microradiography and stereoradiogram. Microradiography produces fine-grain images that can be enlarged considerably. Two radiographs can be combined in a projector to produce a 3D image called a stereoradiogram. Electron microprobes also provide extremely detailed and analytical information of specimens.

In addition to its application in research, X-rays are being used by the industry as a tool for testing objects such as metallic castings without destroying them. Many industrial products are inspected routinely using X-rays to find defects in the products. Ultra soft X-rays are used to determine the authenticity of art work and for art restoration purposes.

The impact of the application of X-rays on healthcare is tremendous. X-ray photography, known as radiography, is used extensively in medicine everyday as a diagnostic tool. People also undergo radiotherapy where X-rays are used to treat and cure conditions such as shrinking tumors and destroying various forms of cancer. Tuberculosis was single-handedly diagnosed by means of radiation when the disease was prevalent. Modern X-ray devices can offer clear views of most of the human anatomy.

9.3.4 Magnetic Resonance Imagery

Magnetic Resonance Imagery or MRI for short was previously known as nuclear magnetic resonance imagery (NMRI). The word, "nuclear" was dropped about 16 years ago to remove fear of nuclear radiation from the minds of the patient, especially since the risk of harmful radiation was actually much less with MRI than with conventional X-rays [3].

MRI is the technique of taking pictures of various parts of the body without the use of x-rays. An MRI is safe for most patients. People who are claustrophobic and who have implanted medical devices such as a eurysim clip in the brain, heart pacemakers, and cochlear implants may not be able to have an MRI. An MRI scanner employs a large and very strong magnet which envelopes the patient. A radio wave antenna is used to send "radio wave signals" to the body and then receive the signals back. These returning signals are converted into pictures by a computer attached to the scanner. MRI is a powerful and versatile tool that generates thin-section images of any part of the body including the heart, arteries and veins, from any angle and any direction without surgical intervention. MRIs can also be used to create "maps" of biochemical compounds within any cross-section of the body. These maps provide valuable biomedical and anatomical information for new knowledge and for early diagnosis of many diseases.

9.3.5 How MRI Works

The human body consists of billions of atoms, the building blocks of all matter. There are many types of atoms in the body, but for the purpose of MRI, we are concerned only with the hydrogen atom. It is an ideal atom for MRI because its nucleus has only a single proton and a large magnetic moment. It means that when placed in a magnetic field, the hydrogen atom has a strong tendency to line up with the direction of the magnetic field.

The principles of MRI take advantage of the magnetic nature of these protons. Once the patient is placed inside the cylindrical magnet, the diagnostic process begins. The process contains five basic steps: (1) the MRI creates a steady state within the body by subjecting the body in a steady magnetic field that is 30,000 times stronger than the earth's magnetic field; (2) A radio frequency (RF) signal is beamed into the magnetic field, which changes the steady-state orientation of the protons; (3) When the RF signal stops, the protons move back to their previously aligned position and release energy; (4) A receiver coil measures the energy released by the disturbed protons. These measurements provide information about the type of tissue in which the protons lie as well as their condition; (5). A computer uses this information to construct an image on a TV screen.

In current medical practices, MRI is primarily used for diagnosing most diseases of the brain and the central nervous system. MRI has the special feature of distinguishing soft tissue in both normal and diseased states. Figure 9.3a shows a person getting an MRI scan, while Figure 9.3b shows an image that is obtained from the scan.

9.3.6 CT

CT and computed axial tomography scans are medical imaging technologies that use both X-rays and computers to produce 3D data of the human body. CT scanners are relatively inexpensive, and are owned by most hospitals. While X-rays traditionally highlight dense body parts such as bones, CT

(a)   (b)

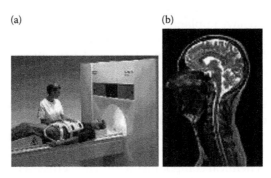

FIGURE 9.3
MRI equipment setup and image.

scanners provide detailed images of the body's soft tissues such as muscle tissues, blood vessels, and even organs like the brain. Unlike conventional X-rays that provide flat 2D images, CT scan provides a cross-section of the body.

A patient undergoing a CT scan is placed on a movable table at the center of a donut-shaped scanner, the size of which is about 2.4 m (8 ft) tall as shown in Figure 9.4. The CT scan system consists of an X-ray source which emits beams of X-rays, an X-ray detector, which monitors the number of X-rays that strike the various parts of its surface, and a computer to analyze and differentiate the detected energy. The X-ray source and detector face each other on the inside of the scanner ring and are mounted so that they rotate around the rim of the scanner. X-ray beams pass through the patient and are recorded on the other side of the detector. As both the source and the detector are in a 360° circle around the patient, X-ray emissions are recorded from many angles. By moving the patients within the scanner, the operation can obtain a series of images called "slices." Each slice represents a slab of the patient's body with a certain thickness (typically 1–10 mm). For CT canners, each slice has 512×512 pixels. Doctors analyze these series of slices to understand the 3D structure of the body.

In order to sharpen an image, patients are sometimes injected with a substance that increases the contrast between different tissues. Patients are also asked to drink a liquid that makes the internal organs more visible in the CT scan. For scanning special parts of the body, such as the chest, abdomen, and pelvis, contrast agents are regularly used.

Since its discovery in the late 1960s and early 1970s by British engineer, Sir Godfrey Newbold Hounsefield and American physicist Allan Macleod

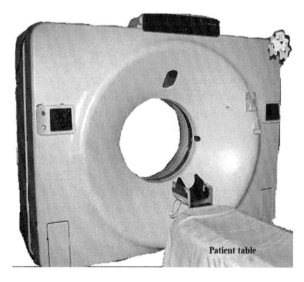

FIGURE 9.4
Typical Computerized Axial Tomography scan machine and equipment setup.

Cormack, the speed at which CT scans are obtained has increased greatly. While the old scanners required 4.5 min of scanning and 1.5 min of computer reconstruction, the new system of CT scanners can create a slice in about a second. The most frequent application of CT is for viewing the tissues of the brain. CT scans of a brain can show if an accident victim has sustained brain injury or if bleeding has occurred in the brain of a stroke victim.

9.4 Software for Making Medical Models

The process of RP enables us to generate a physical model utilizing CAD geometry. The model is made by one of two methods: either a CAD model of the desired object is created first, or a physical object is scanned, and a 3D CAD model is obtained through the reverse engineering process. Because the medical field is an area which greatly benefits from 3DP, various companies have produced software that can be used to convert medical scan data to mathematical representations that are useable by 3DP or CAD systems.

9.4.1 Materialise's Interactive Medical Image Control System and the CT Modeler System

Materialise's interactive medical image control system (MIMICS) is a software system that interfaces between scanner data including CT, MRI, and technical scanners, to modeling software including RP, stereolithography (STL) file format, CAD systems, and finite-element analysis (FEA) [4]. The MIMICS software is an image-processing package with 3D visualization functions that allows CT and MRI images to be segmented, so that they can later become useful physical models.

MIMICS can be used for the purposes of conducting diagnostics, planning surgical procedures, or for teaching and rehearsal purposes. A flexible prototyping system is also included for the building of distinctive segmentation objects. MIMICS enables surgeons and radiologists to control and correct the segmentation of CT scans and MRI scans. What this means is that the object that is being visualized or produced can be defined by the medical staff without any technical knowledge about creating an on-screen 3D visualization of a medical object. Separate software is available that helps to define and calculate the necessary data to build the medical objects created with MIMICS on all RP systems. The MIMICS software is capable of processing vast array of 2D image slices, with the user's computer memory serving as the only restriction.

9.4.2 Visualization Tools

In MIMICS, segmentation masks are used to highlight the regions of interest. Images can be processed and defined with up to 16-colored segmentation masks. Thresholding is the first action performed to create a segmentation mask. A region of interest (ROI) is defined by a range of gray values, with the boundaries of the threshold being the upper and lower range. This is particularly useful in segmenting bone structures within medical CT image data. It is possible to define the segmentation region by two thresholds; thus, this technique can be used for the segmentation of soft tissue in CT images or for the segmentation of multiple structures in MRI images. Additional functions such as the profile function and the histogram function are available to help define distinct thresholds when objects with different materials are scanned. Figure 9.5 shows a human hip along with the pelvis, spine, and femurs, shown on a MIMICS screen. This figure also shows the multiple screens that are available, including a display of a 2D slice, the CT image, and the 3D reconstruction of the pelvic region.

9.4.3 Visualization and Measurements Tools

MIMICS displays CT and MRI image data in several ways, each of which provides unique information. MIMICS divides the screen into three views, the original axial view, and the resliced data making up the coronal and

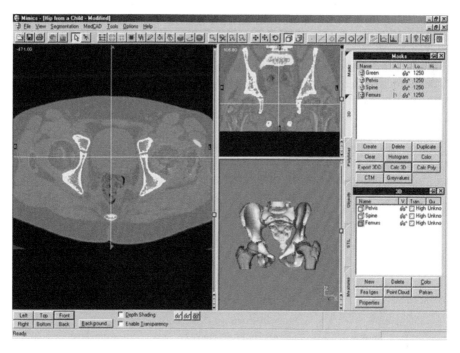

FIGURE 9.5
Example of MIMICS image with multiple threshold masks.

sagittal views. In addition, various options are provided such as zoom factor, contrast enhancement, and the ability to pan across the view and rotate the 3D calculated data. Also provided is a flexible interface to visualize the segmented objects as 3D objects using fast and advanced 3D rendering and shading algorithms.

Measurements can be taken within MIMICS from point to point on 2D slices as well as 3D reconstructions. A profile line displays an intensity profile of the gray values along a user-defined line. Accurate measurements can be made based on the gray value by three methods: the threshold method, the four-point method, and the four-interval method, all of which are ideal for technical CT users.

9.5 Materials for Medical Applications

3DP models used in medical applications can be made with a variety of materials. In some cases, though, these materials must be "medical-grade." Factors that determine what materials to use for an application are much the same factors that one must take into consideration when designing any industrial application. For medical use, practice and simulation models have few constraints as to what kind of material should be used, as the sole purpose of these models is fit check and geometry verification. However, in some cases, 3DP models are used as surgical tools or may be actually implanted to permanently remain in vivo. These types of applications require models with the ability to be sterilized or remain compatible with human tissue. Table 9.1 shows different RP methods, materials used in these methods, and processes of sterilization.

Some other 3DP materials are still in the developmental stages for medical applications, but initial usage of the materials above show that the benefits are quite apparent. As this technology expands, one of the research goals may be to directly copy a patient's anatomical characteristics and fabricate materials that have strengths equivalent to human bones.

TABLE 9.1

RP Materials for Medical Use

3DP Method	Material (Technical Name)	Method of Sterilization
Stereolithography	Somos WaterShed (XC 11122)	Ethylene oxide
Selective laser sintering	DuraForm PA (Nylon 12)	Autoclave
Fused deposition modeling	Biocompatible PolyJet photopolymer (MED610)	Gamma irradiation

9.6 Methodology for Printing Medical Models

Medical models are different from prototype models because medical models are created from individual patient's data. The patient's data come mainly from the MRI and CT scans we have just talked about in the previous section. These models are, sometimes, referred to as stereomodels and medical models represent the 3D sections of an individual patient, generally representing the bone structure including soft tissue, vascular structures, and implants.

A methodology for the printing medical models is shown in Figure 9.6 [5].

The data are usually captured using MRI or CT technology. The data are often represented in a slice format that shows the cross-sections of patient and its tissue types. In order to reconstruct the virtual model of the patient, the image data must be processed to identify and retrieve the relevant materials corresponding to the areas of interest. The contours of each slice are interpolated and joined together to form a 3D model which is key to any 3DP.

Once the CAD model is made, it is then converted to the required STL file. It is the STL file that is used by any 3D printer for making solid models.

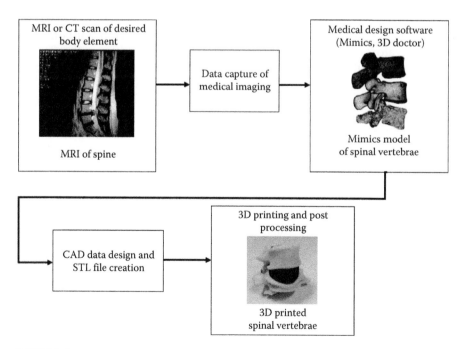

FIGURE 9.6
Methodology flowchart for medical 3D printing.

9.7 Benefits of 3DP in Medicine

There are many benefits of 3DP in medicine. Some of the common benefits are listed below [6]:

1. *Customization and Personalization:* 3DP has facilitated the advancement in customized patient care, facilitating the development of patient-specific treatment plans by printing of patient's anatomy. The customized models help the doctors better prepare for the surgery, rather than relying on the 3D images acquired by MRI or CT scans. Custom-made implants, fixtures, and surgical tools have serious advantages in terms of time taken for the surgery, patient recovery time and the success of the surgery. It is expected that the inkjet-based printers will be used to fabricate drugs and its delivery extensively.

2. *Customization and Collaboration:* Another benefit of 3DP is the customization of design and manufacturing of parts. As a variety of materials are available for making prototypes, they are decreasing the cost of design and manufacturing as well. As a result, more people in both medical and industrial fields are using 3D printer to make new products for personal and industrial use. Another feature of 3DP is the nature of its data which are digital. Parts that designed and converted to STL files can be shared with other educators and researchers to reproduce the parts from the STL file without going through the complex design process. In order to promote open-source sharing of 3D print files, the National Institute of Health has recently established the 3D Print Exchange in 2014.

3. *Increased Productivity:* When RP technology was launched in late 80s, the speed of the machine was not that rapid. Many complex parts used to take several days to print. However, in recent years, there have been tremendous increase in microprocessor speed that resulted in RP speed also. I have a $100,000$ fused deposition modeling machine that was bought 20 years ago. A part that takes 20 h in that machine can be printed in less than 3 h using one of $2,500$ 3DP. Thus, 3DP technology is much faster than conventional computer numerical control (CNC) machines, such as lathe or milling.

4. *Increased Cost-efficiency:* The advantage of 3DP of medical models is its ability to print models very cheaply. As the costs for materials and 3DP machine go down, so does the cost of printing. Many small models can be printed using a few dollars only. Its cost can further be decreased by decreasing the unnecessary resources that are used to print a model.

9.8 3D Printing of a C1 Vertebrate from CT Scan Data

9.8.1 Objectives and Benefits

This project seeks to be the basis for further research in this field and provide an example of how a medical file image can be processed to produce a solid model of a bone for use in the medical field.

The goal of this project was to study medical applications within the field of 3DP, as well as to construct a part through 3DP processes from CT scan data files. This is done to simply show the capability and possibilities 3DP methods have in the medical field. The objectives for project are (1) to acquire either a MRI or CT scan data file of a bone from a medical institution or open-source platform, (2) to process the data file using 3DP-based medical software available and from this to reconstruct a STL file, and (3) to 3DP the STL file using the MakerBot into a physical solid model.

In the case of this project, a CT scan data file of the C1 cervical vertebra of the spine was obtained. The C1 is the first cervical vertebra of the spine, connecting the spine and skull. The CT scan data file was obtained from an open-source platform, and then processed and converted to an STL file. A physical model of this vertebra was then rapid prototyped using the MakerBot.

9.8.2 General Project Process and Software

The task in hand was to convert CT scan data of a Cervical 1 vertebrate into an STL file that could then be used to rapid prototype a model of the body part of interest. There are multiple ways to accomplish this, with varying programs to accomplish each task. Some programs allow more complete manipulation, and can even be used to make easily manipulated CAD models that can be employed in both assemblies and FEA. The most widely used version of this software is MIMICS, which has grown as the industry leader in the application of RP to medical technologies. Unfortunately, as this application becomes more relied upon, more certifications are required for medical use. Gaining these certifications increases the cost of software development and changes the target audience. They are being widely employed within the medical field, and for both of these reasons, many software packages have increased in price significantly. This holds for software that was, at one point, open source.

Due to the inability, a result of a small budget, to use the paid software that is available on the market, an open-source solution was found. While this solution accomplished the original task, the ability to edit was somewhat complicated and the employment of the STL that was created was limited to manufacturing uses. Ideally, a program that would allow some sort of FEA to be conducted would be employed, as having FEA available can be useful in fracture analysis, and models of this type can be used more simply in part development.

There were two software packages used in this project. First, InVesalius 3.0 was used to convert the medical scan data into a STL file. InVesalius 3.0 is a software designed for Windows that was created by a Brazilian team for the Brazilian government. They have since made this software publicly available; however, it is not certified for official medical use. This lack of certification was acceptable for this project, but if being applied to a real person's treatment, certified software would be required. Once the STL of the part of interest is created, MeshLab, an open-source mesh editing software was employed to edit some of the undesired parts away. With more practice, meshes could be augmented to fill holes, however, that was not necessary for this application. MeshLab was used to delete the support structure that was used to hold the vertebrate in place during the scan. Meshlab is a somewhat obtuse program to use, as the method of selection is very limited. This is countered by ensuring a good selection prior to creation of the STL file.

9.8.3 Detailed Process

First, CT scan data was obtained. Originally, the plan was to use the CT scan data of a family member's knee, however, due to the relative complexity of the knee and the incomplete scan folder that was received, this was deemed not useable. Fortunately, CT scan data are available from a professor at the University of Brussels Medical School. From the website, the data for the C1 were downloaded [7]. Next, these data were imported into InVesalius 3.0 using the program's importer. Once imported, the surface of interest must be selected. This was done by choosing the minimum and maximum density to be included on a slider. Below, in Figure 9.7, the slider bar that is used to select the densities to be included. Once selected, it is applied to create a new mask and, if acceptable, a surface is created. The result can be seen as a 3D image in the bottom right-hand corner, as seen in Figure 9.8.

The exported mesh is opened using MeshLab. Using the delete selection tools, located on the upper right menu bar and seen below in Figure 9.9, all extraneous portions of the mesh are deleted.

The unedited mesh can be seen in Figure 9.10. There are parts that are unnecessary as they are not a part of the bone, but were impossible to deselect and still get a good selection of the bone. Using the previously mentioned delete tool, these surfaces were selected and deleted.

After modification, the bone is the only mesh present. The mesh can be further modified to close any holes If necessary, however, in this particular instance, that was not required. The final mesh, seen in Figure 9.11, can then be saved and sent to an RP machine for manufacturing.

9.8.4 3D Printing the Part

The final mesh was converted to STL file and prototyped using MakerBot Replicator 2 at LMU RP lab. Figure 9.12 shows the 3D printed part.

FIGURE 9.7
Surface selection tool.

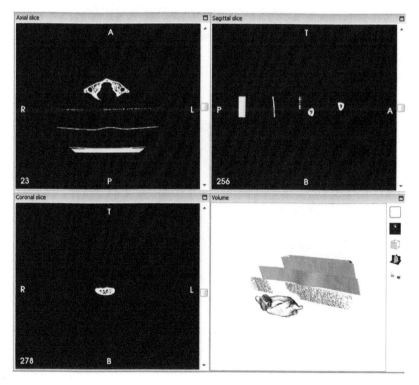

FIGURE 9.8
InVesalius 3.0 View Window. A 3D rendering of the selected surface is seen in the bottom right.

FIGURE 9.9
Meshlab tools used to delete excess portions of the mesh.

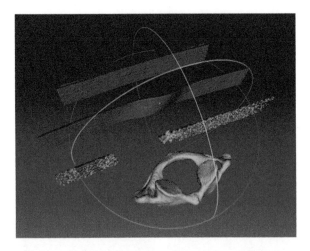

FIGURE 9.10
Unedited mesh immediately after importing to MeshLab.

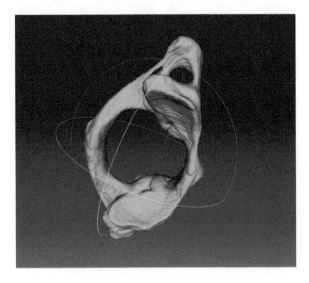

FIGURE 9.11
The final mesh after editing, prior to manufacturing.

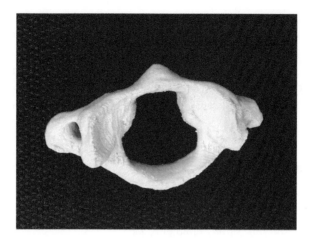

FIGURE 9.12
3DP of the final medical part.

9.9 Summary

There are many different uses of a part like the one created here, such as planning a surgery, developing implants for custom fit, or teaching. In the future, using software that allows the creation of a model this is able to be more completely manipulated and analyzed would offer more information from the same scan. It will undoubtedly help shorten the problem of the donor organ that the society is facing now. All in all, however, being able to produce a physical part from a DCOM format, CT scan provides distinct and substantial advantages to the medical field and can make care provided to students significantly more useful.

9.10 Questions

1. Describe in brief the several areas of applications of 3DP in medicine.
2. What is medical imaging? Why medical imaging is so important for the application of 3DP in medical field.
3. What is an MRI? How does it work?
4. What is a CT scan? How a CT scan is different from an X-ray?
5. What is the most popular software package for medical applications of 3D printer? What are the benefits of the software?

6. A surgeon is about to perform a delicate operation to fix a broken bone of a patient. How does the software help the doctor better plan and execute the operation?

7. What are the special considerations to be taken into account for the use of prototyping materials for medical applications?

8. What is the difference between ABS and medical-grade ABS materials?

References

1. Milwaukee School of Engineering (MSOE), Rapid Prototyping Center, http://www.rpc.msoe.edu/3dmd/3dmd.php, 2002.
2. M. Tabiana, Wilhelm Conrad Röntgen and the discovery of X-rays, https://www.ncbi.nlm.nih.gov/pubmed/8696882, France, 1996.
3. Kaiser Permanente, *Magnetic Resonance Imaging—An Inside Look*, Abi Berger, Bethesda, MD, 2002.
4. Materialise, Belgium, http://www.materialise.com.
5. M. Cooper-White, How 3D printing could end the deadly shortage of donor organs, The Huffington Post, TheHuffingtonPost.com, n.d. Web, April 28, 2015.
6. R. Noorani, *Rapid Prototyping Principles and Applications*, John Wiley & Sons, Hoboken, NJ, 2006.
7. S. Van Sint Jan, Serge, C1, Brussels, http://isbweb.org/data/vsj/, 15 August, 1998, DCOM.

10

How to Select Rapid Prototyping and 3D Printer

10.1 Introduction

The ever-increasing availability and quality of 3D printers and rapid prototyping systems have manufacturers, inventors, artists, and ordinary consumers turning toward 3DP and RP systems as a means of bringing their ideas into the physical world. The 3D printing industry has made several leaps and bounds over the last couple of years and, as a result, 3D printers have a broad range of uses and abilities [1]. However, because of this, there are a lot of different machines and technologies to choose from. Making a decision can be a little overwhelming and intimidating. This chapter will serve as a guide for the various criteria that should be taken into consideration when purchasing an RP and 3DP system. This chapter will also discuss how to obtain access to a 3D printer and RP system, the management and maintenance of the machinery, as well as the development of an expert system. The expert system will help a potential user select an RP or 3DP system by inputting a few parameters of interest. The expert system has been developed using the MATLAB® software package. It is user-friendly and can be modified easily to incorporate new information about the new systems that are coming up.

10.2 Choosing 3D Printer

This section will discuss the different criteria to consider when purchasing a 3D printer such as end product, safety, process, material, quality, and price. The desired final product should drive all other decisions regarding the selections of a 3D printer [2].

10.2.1 Safety and Environment

3D printers are not loud enough to cause hearing loss, and while newer models are silent, some models still give off a considerable amount of noise and noxious fumes. To find a solution to best fit your needs, and budget an online search will turn up some quick and simple solutions to prevent complaints from your neighbors. These solutions work by damping the printer's vibrations, and while some of them will call for expensive materials, some will ask for affordable everyday household items. While the noise level of a 3D printer does not affect the owner's health, the fumes it gives off can be.

Depending on the material and process used, fumes may not be a health issue. However, the fumes are reported to be particularly noxious, and it would be a good idea to take the necessary precautions. Similar to the printer's noise solutions, a quick internet search can also turn up anything from technical level exhaust fans to DIY fume extractors. Most likely you will only have to crack open a window and turn on a few fans.

Another small danger is heat. The material is hot as it comes out, and if the user is not careful, they may burn themselves while trying to handle the printer. As for the electrical needs of the 3D printer, each printer has its requirements; however, a dedicated and spike-free circuit should always be planned for when using this type of equipment.

10.2.2 Process and Material

There are different types of 3D printing and RP processes, and there are also a fair amount of possible materials to choose. Selecting a particular process or material relies heavily on what you want out of the final product. For example, fused deposition modeling (FDM) is more affordable compared to stereolithography, or selective laser sintering (SLS). However, stereolithography and SLS allow for higher quality. SLS also allows the user to print metal objects.

As far as choosing a material to print out the user needs to consider how strong or flexible they want the final product to be. Different materials run different risks, for example, acrylonitryl butadiene styrene (ABS) is better more flexible and has a higher melting point than polylactic acid (PLA). PLA, however, cools quickly preventing deformation as the model cools; ABS takes longer to cool and is susceptible to distortion.

While some 3D printers can use more than one material, some can only handle one. Be sure to research the pros and cons of different printing materials, as well as what material each printer can handle.

10.2.3 Quality of Printer

When investigating the quality of a 3D printer, one needs to note the print speed and printer resolution. The print resolution of a 3D printer tells the buyer the level of detail the printer can handle, while the print speed is how quickly the printer can print. The best way to judge the printers quality is by looking at the models it produces. Most companies offer print samples, and a quick internet search can also show you different finished products of the printer. It is important to note that the output of the 3D printer is dependent on the hardware and the materials selected. The quality of a prototype is measured by dimensional accuracy, surface finish, and strength of the prototype.

10.2.4 Price

While significantly cheaper than when 3D printers were first available, the technology can still be pretty costly. After all, if you are going to invest more than $350 (minimum) for just the printer alone, you want to get the one you will work best for you. The price of 3D printers can easily surpass $3000. The average price for a consumer 3D printer is around $400–$1000. In addition to the printer, you may also need to purchase additional materials for noise reduction, air circulation, and print material. Modeling software (potentially for free), postcure oven (if necessary), cleaning materials, and storage supplies may also need to be purchased. 3D printing filament costs an average of $30.00 per kg.

It is important for the consumer to take into consideration any hidden costs, such as operation, postprocessing, and building modifications. Depending on the 3D printer or RP system purchased, electrical, plumbing and heating, ventilation, and air conditioning (HVAC) work may need to be done to ensure the safe use of the 3DP or RP system. While this precaution is typically for systems used by the industries even smaller 3D printers may require extra ventilation.

The price of traditional RP system is still very high. Some of the models from Stratasys and 3D Systems can cost anywhere from $20,000 to a $1,000,000.

However, these RP systems have many desirable advantages such as bigger build volume, faster speed, versatility of materials, and the strength and accuracy of parts built.

10.3 Operating Issues

3DP technology is still relatively new, and you are likely to run into problems eventually, and depending on the complexity of the printer and the problem itself, you may or may not be able to resolve the issue yourself. This section will discuss what the individual needs, might be, to maintain their 3D printer properly for as long as possible.

10.3.1 Learn to Use It

If someone is just beginning to dip their feet into 3D printing and has not used a 3D printer before, they will not be able to figure out how to properly use the printer on their own, like they would an ink jet printer. A new 3D printer owner should try to place several days aside for learning how to use the system as well as several weeks for hands-on practice. Usually, the machine vendor can be contacted to train the people responsible for the operation and maintenance of the machinery. Even someone who is no stranger to the world of 3D printing may have trouble with setup.

10.3.2 Learn to Maintain It

Once the printer has been properly set up and calibrated, it will need to be regularly serviced to keep it running smoothly. Vendors usually offer these services, and can even give you advice, and tips on how to get the most out of your printer. If it is a simple problem, you may even be able to get assistance right from their website. The bottom line is that the vendor knows how to take care of the system, and they are not going to hide that information from their customers.

Online communities are also very beneficial for the care of a 3D printer. Many communities dedicate themselves to 3D printing, and a majority pertain to a particular type of printer. These communities contain people who are often very passionate about their work and have come up with several tips and tricks that improve the quality of the 3D printer. These online forums can also serve to be very beneficial in gathering information about select printers.

Not all problems are complicated enough to call in the experts. Sometimes, all someone needs to do is untangle the filament or level the print bed. Filament should be properly stored in a cool, dry, and clean environment. The printer's extruder is easily clogged, so it is important to keep filament in its ideal condition. The filament can also become tangled easily if the user is not careful when loading and unloading the filament. However, should the filament become tangled there is no need to panic, simply unwind the filament, detangle it, and rewind it back onto the spool and reload.

If the printer's extruder is not extruding the proper amount of filament, it is most likely clogged. Simply unloading and reloading the filament will usually fix the problem. It should be noted that the extruder should be inspected for any clogs before printing.

The print bed also has a large impact on the quality of the print. If the bed shifts during print, or is not level, layers in the print will be moved resulting in a failed print. To combat bed warping the user can either put some glue down on the bed directly or cover the bed with masking tape. A 3D printer will also give you instructions on how to level the print bed. Instructions are located in the printer's settings.

10.4 Accessing 3DP and RP Systems

Investing in 3D printing and RP systems can be expensive. Cheap 3D printers are available, but unfortunately, most of them have poor print quality. Poor quality may be okay for some, but for those who need higher precision or quality material, the affordable printer will not suffice. Fortunately, for those who need access to printers outside their price range, outsourcing is also an option for obtaining more affordable prints.

10.4.1 Service Bureau

A service bureau is a company that owns and operates one or more RP machine and manufactures the prototypes of other corporations or private user for a fee. Service bureaus also offer other services related to the manufacturing process such as, but not limited to, concept developments, computer-aided design (CAD), data translation, prototyping, rapid tooling, casting, and reverse engineering [3]. While service bureaus do charge for these services, their fees are a cheaper alternative to purchasing and implementing an RP system.

Because investing in 3DP and RP systems is a pricey endeavor it is important to make the right decision in the final purchase. If someone invests in the wrong printers they can easily find themselves losing hundreds, if not thousands of dollars. At a service bureau, the customer can easily experiment with various RP technologies before making a purchase decision.

Depending on the kind of work one wants to do with the 3D printer, it may be in their best interest to outsource to a service bureau, as 3DP and RP technologies as the investment in those technologies do not always pay back.

10.4.2 Popular Service Providers

Some of the popular service providers are mentioned below. These providers serve the needs of designers, hobbyists, or anybody who wants to print something without purchasing a 3D printer. People can upload their design files to the websites of the companies, get a free estimate of the cost and order the print. It is that simple. These companies have a variety of printers that can use different types of materials to print the part of your choice. Some of the providers are as follows:

Shapeways: http://shapeways.com

Ponoko: http://ponoko.com

Sculpteo: http://sculpteo.com

i.matreialise: http://i.materialise.com

10.4.3 Consortia

There are two types of Consortia: university–industry consortia, and industry–industry consortia [4]. The more common of the two is the university–industry consortia. This kind of consortia involves the cooperation of universities and industry to advance RP. Industry–industry consortia include only industry, and together they look for ways to improve RP applications as it relates to their particular industry. One of the benefits of consortia affiliation is the monthly networking sessions and workshops. In these networking sessions, companies with new RP machines can discuss and resolve some of their problems with experienced users.

Since 3D printers are inexpensive compared to RP systems, the idea of service bureau and consortium is more relevant to the implementation of RP system.

10.4.4 Build Your Own

One alternative to purchasing a completed 3D printer is building your own. There are 3D printing kits available if this is the desired route. This option also allows the user a little more freedom when tinkering with the technology. However, building your 3D printer takes a considerable amount of time, and, if the user wants to begin printing right away, this may not be the best choice. In that case, selecting a 3DP from the marketplace that suits your needs is the right choice. Chapter 3 provides you with the knowledge of designing and building a 3D printer.

10.5 Development of an Expert System

The ever-increasing demand for quality products and lower costs has manufacturers turning to the new technology of RP. The successful implementation of RP depends on the appropriate selection of the prototyping system and effective planning. The availability of over twenty different companies with wide ranging capabilities and specifications creates a problem of selecting the right RP system for industry as well as academic institutions. The problem is more compounded by the emergence of 3D printers, which are smaller than conventional RP systems and sell for under $5,000. This section presents an expert system that helps the user choose an RP system such as an stereolithography apparatus (SLA) machine from 3D Systems and an FDM machine from Stratasys as well as 3D printers such as MakerBot-2 and Printerbot Simple. The program allows the user to choose a particular system based on the inputs of the user to the expert system. The expert system has been developed using rule-based features of the MATLAB, a popular software used mostly for educational purposes. This is the first expert system that helps user select both a traditional RP as well as 3D printer systems. The architecture of the expert system is such that it can be easily expanded to take into considerations the emergence of newly developed RP and 3D printers.

10.5.1 Introduction

The goal of this section is to describe and discuss the development of an expert system for the selection of an RP and 3D printing system for both industry and academia. The specific objectives of the section are to discuss (1) the need for an RP system for product development and other applications, (2) the need for an expert system to be used as a selector for both the traditional RP systems and 3D printers, (3) to discuss the architecture of the expert system, and (4) to show the application of the expert system for the RP and 3D printers selection processes.

10.5.2 The Need for RP

RP is a process used for fabricating solid prototypes from a CAD data file [5]. RP can be applied in diverse fields of the industry, with prototypes used for form, fit and function, design verification, early detection of error and reduction of waste [6]. Design engineers around the world use RP to pre-estimate product characteristics such as shape, manufacturability, and surface finish. The technology is used to build physical models, prototypes, patterns, tooling components, and production parts in plastics, metal, ceramic, glass, and composite materials [7]. According to Wohlers Report 2014, there are over 20 companies around the world working on different RP processes. RP systems can be classified in a variety of ways depending on the physics of the process, the source of energy, type of material, size of the prototypes, etc. Most RP systems are classified as solid-based, liquid-based, and powder-based by the determination of the initial form of the raw materials. Even under any one system such as solid-based systems, there are many types including Stratasys' FDM, Helysis's laminated object modeling, KIRA's selective adhesive, and hot pass. Table 10.1 shows the RP systems with different work volume, materials, cost, etc., from just one company and one type of additive manufacturing.

The selection of RP systems in earlier days was based on the benchmarking that was done either by manufacturers or by independent researchers. This benchmarking study compared the relative strengths and weaknesses of each company based on the quality, reliability, part accuracy, tensile strength, surface finish, etc. More recently, researchers have been developing computer programs that help people from industry and academia select a particular RP system. However, there is a new challenge on the horizon. There is a revolution that started in 1982 but did not proliferate much for years. This new revolution of prototyping system is called 3D printers [8]. 3D printers are smaller in size, much less expensive, and can be built by an individual or a group of researchers using only a few thousand dollars. Most conventional RP systems still cost $30 K to $500 K. Despite the growing movement, there has been no report of any research on the selection of 3D printers. Before purchasing an RP system, two practical but common sense questions should be considered. First, should a particular RP process be selected over other prototyping process for a given type of product? Second, which RP process should be chosen based on which aspect of the technology will be pursued by this company? A number of criteria should be evaluated

TABLE 10.1

Stratasys' Material Jetting Printers

Model Name	Build Volume mm (in.)	Material	Approximate Price × 1000
Objet 1000	$1000 \times 800 \times 500$ $(39.4 \times 31.5 \times 19.7)$	Proprietary acrylate photopolymers, over 100 materials and digital materials	$600
Objet 500 Connex3	$500 \times 400 \times 200$ $(19.7 \times 15.7 \times 7.9)$	Proprietary acrylate photopolymers, over 500 digital materials and colors	$100
Connex500	$500 \times 400 \times 200$ $(19.7 \times 15.7 \times 7.9)$	Proprietary acrylate Photopolymers, over 100 materials and digital materials	$240
Connex350	$350 \times 350 \times 200$ $(13.8 \times 13.8 \times 7.9)$	Proprietary acrylate photopolymers, over 100 materials and digital materials	$200
Objet 260 Connex	$260 \times 260 \times 200$ $(10.2 \times 10.2 \times 7.9)$	Proprietary acrylate photopolymers, over 100 materials and digital materials	$179
Eden 500V	$500 \times 400 \times 200$ $(19.7 \times 15.7 \times 7.9)$	Proprietary acrylate photopolymers, biomaterials	$174

when considering the first question. These criteria depend on the product application and may include: cost trade-off, cycle time (throughput), accuracy of the prototyped parts including tolerance and surface finish, material properties (material selection), size of the part, strength of the part, etc.

10.5.3 Need for an Expert System

Selecting the right RP system is absolutely essential for any company that has identified a need for this equipment. However, the success of the implementation of the RP system depends largely on the people who operate and maintain the RP machine and its support equipment. There is a myth about the RP machines that "the systems are as simple as pushing a button." This is not so. As with any complex system, it can often take a company several weeks to over a month to get the machine running properly. Based on a positive experience with their first machine, many companies have bought several additional machines for higher productivity as well as reduced cost and wastage. RP is a new technology, and as such, it takes a great deal of experimentation and practice to make quality prototypes that will be useful to the end-user.

An expert system is one of the elements of artificial intelligence concerned with the use of a computer in tasks that are normally considered to require knowledge perception, reasoning, learning understanding and similar cognitive ability. An expert system is an intelligent computer that uses knowledge and inference procedures to solve problems that are difficult enough to require significant human expertise for their solution [9].

The essential features of an expert system are as follows:

1. A knowledge base of domain facts and heuristics associated with the problem.
2. An inference procedure for the utilization of the problem.
3. And a working memory or global database for keeping track of the problem status.

An expert system consists of a knowledge base and an inference engine. The knowledge base is the representation of knowledge and the inference engine is the plausible reasoning. Figure 10.1 shows the architecture of an expert system.

10.5.4 Literature Review

Because of the varieties of RP systems with wide ranging capabilities, researchers have been selecting a prototype based on the benchmarking studies. In a benchmarking study, a benchmark part is created using a CAD system with features that will test the capabilities of the rapid prototype machines. Using a stereolithography file, the benchmark part is made.

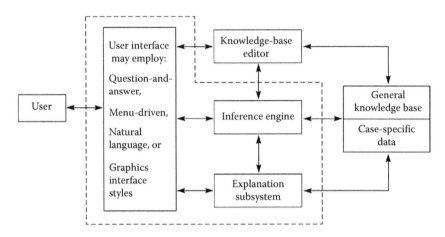

FIGURE 10.1
Architecture of an expert system.

The part is then measured and tested for its accuracy, surface finish, and other desirable features. While benchmarking is a helpful method, it is time-consuming and expensive [10]. Recently, researchers have developed computer programs that help select a particular RP machine based on the user providing answers to specific questions [11]. The computer program provides the recommendation and suggestions. This method of selecting a prototyping machine is quicker and more interactive than the benchmarking method. However, this computer program does not involve any expert system.

The first attempt to develop a computer program for selecting an RP machine was made by Hornberger et al. at Santa Clara University in 1993 [12]. The program was based on the RP machines available at that time such as 3D System's SLA machine, Stratasys' FDM machines, etc. In 1995, Muller et al. [13] at Bremen Institute of Industrial Technology and Applied Work Science developed the first RP system selector. This program uses the relational database management system MS Access and is popular for the RP community. The program however has its limitations. Initially, it was supposed to select a machine based on the combination of factors such as solid, hollow, thin walled, etc., and a particular RP process.

In 1996, Phillipson [14] from Arizona University developed the RP Advisor that also used the MS Access. This program was compared to that of Muller using Japanese methodology of quality function deployment. As the technology expanded and the revolution of 3D printers took off, it became more difficult to select a prototyping machine that met the needs of a customer or a company. An expert system is an intelligent system that captures the human knowledge and experiences in a computer program. The system can be used to select a prototyping system and to solve other problems such as process planning, process parameter optimization, and tool design. Expert systems have been developed for part orientation and concurrent design in RP. Masood and Soo [15] developed the IRIS intelligent RP selector to assist RP users in the industry and academia to quickly select the RP system that met their requirements. The program incorporates most of the models from the United States of America, Japan, Germany, Israel, and other countries. This program is open ended and can incorporate future prototyping systems.

This section presents the development of an expert system to select the appropriate RP machine that fulfills the requirements of the user. As previously mentioned, all benchmarking or computer-based expert systems have addressed the issue of selecting conventional RP systems. However, there is a tremendous interest in the 3D Printing systems that have proliferated in recent years. No program has addressed the issue of selecting a 3D printer system.

This section will present an expert system that can be used to select both conventional prototyping machines as well as 3D printing machines. The system has been developed using the popular MATLAB software and uses a multi paradigm rule-based system.

10.5.5 Creating the System

The development of an Expert System for the selection of an RP system and 3D printers involved a process of three major steps. First, factors were selected that affected the selection of the systems. These factors were then prioritized and a multicriteria decision analysis was used to optimize the selection. Finally, a MATLAB program was developed that acts as the medium for the expert system. The MATLAB code aids in the selection process by asking the user for various inputs. It then outputs which system the user should purchase. The final product will be an expert system that is user-friendly. Upon the completion of the MATLAB code, the researchers will revise the factor selection, systems involved, and decision process in order to increase the scope of the expert system.

10.5.6 Selection of Factors

The selection of factors was based on the information provided by various companies regarding the specifications of various systems listed on their websites. In order to determine which factors were relevant for the selection process, the systems themselves first had to be selected. The selected systems and their specifications can be seen in Tables 10.2 and 10.3.

All these parameters play a large role in the selection of an RP system and a 3D printer and this information was available on the companies' specification tables of their various systems. These systems were selected based on their reasonable prices and availability to the general population.

TABLE 10.2

3D Printer Specifications

3D Printers	Printerbot Simple 2016	Bukito	Ultimaker 2+	MakerBot Replicator 2×	Cube 2	Formlabs 1+	Lulzbot Taz 6
Platform length (cm)	20.3	14	22.3	24.6	13.97	12.5	28
Platform width (cm)	15	15	22.3	15.2	13.97	12.5	28
Platform height (cm)	20.3	12.5	20.5	15.5	13.97	16.5	25
Price (USD)	999	799	2499	2499	1299	1999	2500
Roughness (resolution) (um)	50	50	20	100	67	25	80
System length (cm)	48.26	30.48	49.4	49	25.4	30	66
System width (cm)	40.64	30.48	34.2	42	25.4	28	52
System height (cm)	50.8	30.48	58.8	53.1	33	45	52
Weight (kg)	7.25	2.72	11.3	12.6	4.3	8	11
Speed (mm/s)	80	100	30	135	20	20	200

TABLE 10.3

RP System Specifications

RP Systems	EOS M 100	Stratasys Fortus 450 mc	Stratasys Object 500 Connex3	EnvisionTec ULTRA 3SP	3DS ProJet 6000 SD	Stratasys Fortus 380 mc	Optomec LENS 450
Platform length (cm)	10	40.6	49	26.6	25	35.5	10
Platform width (cm)	10	35.5	39	17.5	25	30.5	10
Platform height (cm)	9.5	40.6	20	19.3	25	30.5	10
Price (USD) X1000	100	100	100	50	200	100	299
Roughness (resolution) (µm)	40	200	20	100	100	200	25
System length (cm)	80	129.5	140	74	78.7	129.5	100
System width (cm)	95	90.2	126	76	73.7	90.2	100
System height (cm)	225	198.4	110	117	182.9	198.4	150
Weight (kg)	580	680	430	90	181	680	200

10.5.7 Prioritization and Selection

Once the factors and systems were selected, the individual factors had to be weighed. These weights are inputted by the user in a scale from one (lowest) to ten (highest). For example, the user can weigh "price" to be 10 while "roughness" to be 5; meaning that the surface roughness worth half the price. The multicriteria decision engine was then determined. This engine needed to be able to quickly determine which system best fits the user needs; this was done through the use of a point system. Once the user inputs their information, the system will assign one point to the system that fits the user requirements for each parameter that fits the system specifications. This one point assigned is then multiplied by the weight inputted by the user. The system with the most points gets selected as the best fit for the user. For example, if the user inputs a maximum price of $50,000, only the EnvisionTec ULTRA 3SP will receive the points. Additionally, if the user inputs a minimum platform length of at least 30 cm, the Stratasys Fortus 450 mc and 380 mc, and Stratasys Object 500 connex3 will all receive the points; however, the other systems will not receive a point. Figure 10.2 shows a diagram of the process.

10.5.8 Development of MATLAB Program

A medium for the expert system was needed in order to provide the user with some way of relaying their requirements to the multicriteria decision engine. MATLAB was selected as this medium based on the proficiency of the programmer with this programming language and its ease of access at Loyola Marymount University. The program was carefully developed to follow the multicriteria decision engine. The program assigns points to the systems that fit the user requirements and outputs the system that best fits the user needs. What follows are small pieces of the program used to compose the expert system.

The following is the user input section of the code:

l = input ("Please enter the minimum build platform length desired (cm): ");

ll = input ("In scale of 1 (lowest) to 10 (highest), what weight would you give the previous parameter?");

w = input ("Please enter the minimum build platform width desired (cm): ");

ww = input ("In scale of 1 (lowest) to 10 (highest), what weight would you give the previous parameter?");

h = input ("Please enter the minimum build platform height desired (cm): ");

hh = input ("In scale of 1 (lowest) to 10 (highest), what weight would you give the previous parameter?");

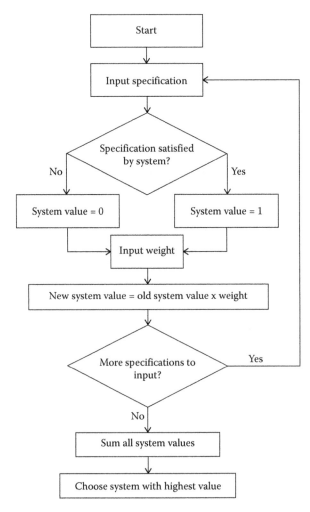

FIGURE 10.2
Flow diagram; decision-making process of the expert system.

p = input ("Please enter the maximum price (USD): ");

pp = input ("In scale of 1 (lowest) to 10 (highest), what weight would you give the previous parameter?");

r = input ("Please enter the maximum desired surface roughness (resolution) (um): ");

rr = input ("In scale of 1 (lowest) to 10 (highest), what weight would you give the previous parameter?");

L = input ("Please enter the maximum desired system physical length (cm): ");

LL = input ("In scale of 1 (lowest) to 10 (highest), what weight would you give the previous parameter?");

W = input ("Please enter the maximum desired system physical width (cm):");

WW = input ("In scale of 1 (lowest) to 10 (highest), what weight would you give the previous parameter?");

H = input ("Please enter the maximum desired system physical height (cm):");

HH = input ("In scale of 1 (lowest) to 10 (highest), what weight would you give the previous parameter?");

z = input ("Please enter the maximum desired system weight (kg): ");

zz = input ("In scale of 1 (lowest) to 10 (highest), what weight would you give the previous parameter?");

s = input ("Please enter the minimum desired speed of system (mm/sec): ");

ss = input ("In scale of 1 (lowest) to 10 (highest), what weight would you give the previous parameter?");

The following is the database that the program uses to determine which factors fit the user requirements. This is the information for one system (The Printerbot Simple).

al = 20.3;
aw = 15;
ah = 20.3;
ap = 999;
ar = 50;
aL = 48.26;
aW = 40.64;
aH = 50.8;
az = 7.25;
as = 80;

Once the program assigns points to each system and multiply them by their weights, these points are put into the seven different parameters below (a–g) each representing a different RP system. The program then selects that holds the most value, therefore, selecting the system that best fits the user requirements.

X = [a b c d e f g],
F = max(X).

10.5.9 Running the Program

Running the program is quite simple. The user will initiate the code in MATLAB, thus prompting the user to answer a variety of questions. The user will then answer the questions to the best of their abilities. This will initiate the logic sequence of the code. It will take a few moments to analyze the data input by the user and run it through the system. After processing the data input by the user, the program simply outputs the system that would best fit the requirements of the user along with a brief explanation about the system.

Example

Consider a user is looking of for a 3D printer with particular specifications. Initially, the user has to specify whether the user is looking for an RP system or a 3DP by answering the question shown in Figure 10.3. Next, he should input all of his specifications and weights for each parameter as shown in Figure 10.4. Figure 10.4 shows a screenshot of the MATLAB software with the input parameters by the user.

Once the user entered all parameters and weights, the program gives points with the corresponding weight to each system if they satisfy the requirements. Table 10.4 shows how the program assigns points to each system. The system with highest points will be selected as the best to fit the user requirements.

```
Are you looking for a Rapid Prototyping system or a 3D Printer? For RP, please enter 1. For 3DP, please enter 2):
```

FIGURE 10.3
User inputs the selection of either RP system or 3D printer.

```
Please enter the minimum build platform length desired (cm): 15
In scale of 1 (lowest) to 10 (highest), what weight would you give the previous parameter? 9
Please enter the minimum build platform width desired (cm): 15
In scale of 1 (lowest) to 10 (highest), what weight would you give the previous parameter? 9
Please enter the minimum build platform height desired (cm): 20
In scale of 1 (lowest) to 10 (highest), what weight would you give the previous parameter? 10
Please enter the maximum price (USD): 2000
In scale of 1 (lowest) to 10 (highest), what weight would you give the previous parameter? 5
Please enter the maximum desired surface roughness (resolution) (um): 50
In scale of 1 (lowest) to 10 (highest), what weight would you give the previous parameter? 8
Please enter the maximum desired system physical length (cm): 50
In scale of 1 (lowest) to 10 (highest), what weight would you give the previous parameter? 7
Please enter the maximum desired system physical width (cm): 50
In scale of 1 (lowest) to 10 (highest), what weight would you give the previous parameter? 7
Please enter the maximum desired system physical height (cm): 50
In scale of 1 (lowest) to 10 (highest), what weight would you give the previous parameter? 7
Please enter the maximum desired system weight (kg): 10
In scale of 1 (lowest) to 10 (highest), what weight would you give the previous parameter? 3
Please enter the minimum desired speed of system (mm/sec): 40
In scale of 1 (lowest) to 10 (highest), what weight would you give the previous parameter? 6
```

FIGURE 10.4
Inputs by user.

TABLE 10.4

Program Methodology

Parameter	Specification	Weight	Printerbot Simple 2016	Bukito	Ultimaker 2+	MakerBot Replicator 2x	Cube 2	Formlabs 1+	Lulzbot Taz 6
						3D Printers			
Platform length (cm)	15	9	$1 \times 9 = 9$	$0 \times 9 = 0$	$1 \times 9 = 9$	$1 \times 9 = 9$	$0 \times 9 = 0$	$0 \times 9 = 0$	$1 \times 9 = 9$
Platform width (cm)	15	9	$1 \times 9 = 9$	$1 \times 9 = 9$	$1 \times 9 = 9$	$1 \times 9 = 9$	$0 \times 9 = 0$	$0 \times 9 = 0$	$1 \times 9 = 9$
Platform height (cm)	20	10	$1 \times 10 = 10$	$0 \times 10 = 0$	$1 \times 10 = 10$	$0 \times 10 = 0$	$0 \times 10 = 0$	$0 \times 10 = 0$	$1 \times 10 = 10$
Price (USD)	2000	5	$1 \times 5 = 5$	$1 \times 5 = 5$	$0 \times 5 = 0$	$0 \times 5 = 0$	$1 \times 5 = 5$	$1 \times 5 = 5$	$0 \times 5 = 0$
Roughness (um)	50	8	$1 \times 8 = 8$	$1 \times 8 = 8$	$1 \times 8 = 8$	$0 \times 8 = 0$	$0 \times 8 = 0$	$1 \times 8 = 8$	$0 \times 8 = 0$
System length (cm)	50	7	$1 \times 7 = 7$	$1 \times 7 = 7$	$1 \times 7 = 7$	$1 \times 7 = 7$	$1 \times 7 = 7$	$1 \times 7 = 7$	$0 \times 7 = 0$
System width (cm)	50	7	$1 \times 7 = 7$	$1 \times 7 = 7$	$1 \times 7 = 7$	$1 \times 7 = 7$	$1 \times 7 = 7$	$1 \times 7 = 7$	$0 \times 7 = 0$
System height (cm)	50	7	$0 \times 7 = 0$	$1 \times 7 = 7$	$0 \times 7 = 0$	$0 \times 7 = 0$	$1 \times 7 = 7$	$1 \times 7 = 7$	$0 \times 7 = 0$
Weight (kg)	10	3	$1 \times 3 = 3$	$1 \times 3 = 3$	$0 \times 3 = 0$	$0 \times 3 = 0$	$1 \times 3 = 3$	$1 \times 3 = 3$	$0 \times 3 = 0$
Speed (mm/s)	40	6	$1 \times 6 = 6$	$1 \times 6 = 6$	$0 \times 6 = 0$	$1 \times 6 = 6$	$0 \times 6 = 0$	$0 \times 6 = 0$	$1 \times 6 = 6$
Total			64	52	50	38	29	37	34
Rank			1	2	3	4	7	5	6

```
The Printerbot Simple 2016 best fits your needs.
The specification(s) that was/were not fulfilled is/are:
system height

Do you want to see the second best (For yes, please enter 1. For no, please enter 2)?
```

FIGURE 10.5
MATLAB outputs the best choice.

```
Do you want to see the second best (For yes, please enter 1. For no, please enter 2)? 1

The Bukito comes in the second place.
The specification(s) that was/were not fulfilled by the second best system is/are:
platform build length
platform build height
```

FIGURE 10.6
MATLAB outputs the second best choice.

From Table 10.4, it can be noticed that Printerbot Simple 2016 3D printer ranked first with 64 points. In MATLAB, once the user clicks "Enter" after inputting specifications and weights, the software should select Printerbot Simple 2015 3D printer as the best to fit the user's needs. The program also demonstrates all of the specifications that were not fulfilled, or satisfied. In this example, the system height did not match the specifications of the Printerbot Simple 2016 3D printer; therefore, zero points for this factor (Table 10.4). Figure 10.5 shows MATLAB output.

From Figure 10.5, it can be noticed that the MATLAB program also asks if the user wants to see the second best option for him by entering the value 1. If the user enters 1, MATLAB will show the second best 3D printer that fits the user needs and the specifications that were not fulfilled by the input information. In this example, as it can be observed from Table 10.4, the Bukito 3D printer ranked second with 52 points. The specification failures were platform build length, and platform build height. Figure 10.6 shows the MATLAB output when the user input 1.

10.5.10 Limitations and Future Improvements of the Program

The MATLAB code is quite well-suited to provide the user with a system that best fits their needs. It does this quickly and has seven different systems it can choose from. This gives the user a decent range of systems, therefore, allowing the program to select a system that best fits the user requirements. However, in order to better improve the program, additional systems can be added to give the user a large range of systems and the program a method of truly selecting the best option for the user. Additionally, there could be an auto-update feature to the program that adds printers as they are released to the market. This would prevent the program from becoming obsolete and allow the widespread use of the program. Also, the program could be rerun if the user does not agree with a certain system output. When rerun, the system that they did not believe fit their requirements would be removed from the selection engine and a new system would be selected. Currently, the program is split to both an RP selection engine and a 3D printer selection engine. As the program develops, these will be combined so the user does not have to run two different programs if looking for both an RP system and 3D printer.

10.5.11 Conclusion

The expert system developed in this section is an interactive program that helps a potential buyer in the industry and academia to select a conventional RP system or 3D printer from a wide selection of commercially available systems. The user can use a wide range of selection criteria to select a system that meets his/her requirements. Another feature of the program is that it is open ended so that the program can be easily modified to take into consideration the availability of new products in the future. Additionally, the program is user-friendly, easy to use, and has a potential to be used on a wide scale.

10.6 Present and Future Trends

Because modern RP techniques and conventional prototyping techniques such as computer numerical control (CNC) machining are reducing lead time design and manufacturing industries are using RP to increase the product development at increasing rates. More than 1 million prototypes are created each year and more than a million engineering workstations have been installed, this trend is expected to continue of the course of the next several years. Advancement in 3D printing technology is focused on improving the quality of the end product for direct use by the customer rather than a model of a prototype.

Even though 3D printing is available to most commercial users, work is being done to make it even more accessible. Tactile vision graphics Inc. is currently trying to create the world's first braille friendly 3D printer, making the technology available to the visually impaired.

As 3D printing technology continues to advance so will the capabilities of the technologies it can produce. For example, the world's first 3D printed plane was unveiled in June 2016. Researchers hope 1 day to be able to print biological materials such as heart tissue and complex organs. Companies like Collplant are working to create the collagen-based bioink. While researchers are currently unable to successfully print functioning organs they are extremely close to doing so. In addition to this, many amputees are now able to get affordable prosthetic limbs. Dentists can 3D print patient's mouths for better fitting molds. Doctors are also better able to plan for surgery with 3D printed models of a patient's heart.

10.7 Summary

In this chapter, we have presented the guidelines one should follow for implementing RP and managing the related issues. RP is still relatively new and can be an expensive investment. Because of this, purchasing decisions should be based on sound criteria, such as cost trade-off, cycle time, accuracy and surface finish of prototypes, size, and strength of the part. Considerations for facility planning should also be taken into consideration

Later, this chapter discussed the operating issues involved with RP systems such as equipment cost, building cost, training, and maintenance cost. Since RP is not a technology in itself, it draws on other technologies such as CAD/CAM, CNC machining, etc. Once RP is properly integrated into a company or work place, the user will experience many direct and indirect benefits.

Purchasing an RP system may not be feasible for every company, especially, for a small company or individual. Service bureaus and consortia can provide the RP needs of a company. We also discussed the criteria of choosing a service bureau or consortia. This chapter also discussed present and future trends of the technology involved in the RP system.

Based on the above discussions, it can be safely said that an RP and its associated technologies are having a profound impact on the way companies produce models and prototype parts.

10.8 Questions

1. What is the role of 3D Printer for product development? What are the options a company has to implement 3DP in their workplace?
2. Selection of a 3D printer is not an easy job. What are the criteria a company need to evaluate for selecting a particular type of rapid prototyping or 3D printing machine?
3. What kind of material you would use and why?
4. Are all service bureaus the same? Are there any hidden costs? If so, what are they?
5. What is an expert system? Why do you need an expert system for the selection of RP and 3DP?
6. How do you build an expert system? What kind of software is useful for building an expert system and why?
7. SRZ Corporation is planning to purchase a rapid prototyping system at a price of $75,000. The first-year operation and maintenance cost

is $15,000. The Company is eligible for $10,000 tax credit from the federal government under its technology investment program. The RP system replaces 2 design and analysis workers. The hourly rate for the worker is $25 including fringe benefits. Determine the payback period for the company for its investment.

References

1. C. Kai and L.K. Fai, *3D Printing and Additive Manufacturing, Principles and Applications*, 4th Edition. World Scientific, Singapore, 2015.
2. T. Grimm, *User's Guide to Rapid Prototyping*. Society of Manufacturing Engineers, Dearborn, MI, 2004.
3. C. Coward, *Idiot's Guide As Easy As It Gets 3D Printing*. Alpha by Penguin Group, New York, 2015.
4. J.F. Kelly, *3D Printing: Build Your Own 3D Printer and Print Your Own 3D Objects*, Indianapolis, IN.
5. R. Noorani, *Rapid Prototyping: Principles and Applications*. John Wiley & Sons, Englewood, NJ, 2006.
6. M.P. Groover, *Automation, Production Systems, and Computer Integrated Manufacturing*. Prentice-Hall, Englewood Cliffs, NJ, 2001.
7. O. Es-Said, R. Noorani, M. Mendelson, J. Foyos, and R. Marloth, Effect of layer orientation on mechanical properties of rapid prototyped samples, *Materials and Manufacturing Processes*, Vol. 15, No. 1, pp. 107–122, 2000.
8. T. Wohlers, Wholers report 2014: 3D printing and additive manufacturing state of the industry Wholers, Fort Collins, CO.
9. B. Buchanan and E. Shortcliffe, *Rule-Based Expert Systems*. Addison-Wesley, Reading, MA, 1984.
10. L.D. Schmidt, A benchmarking comparison of commercial techniques in rapid prototyping. *Rapid Prototyping and Manufacturing Conference*, Dearborn, MI, 1994.
11. D. Frank and G. Fadel, Expert system based selection of the preferred direction of build for rapid prototyping processes, *Journal of Intelligent Manufacturing*, Vol. 6, No. 5, pp. 339–345, 1995.
12. L.E. Hornberger et al., *Rapid Prototyping Program*. Santa Clara University, Santa Clara, CA, 1993.
13. H. Muller et al., Computer-based rapid prototyping system selection and support. University of Bremen, 1995.
14. D.K. Phillipson, Rapid Prototyping machine selection program. M.Sc. thesis, Arizona State University, 1996.
15. S. Masood and A. Soo, A Rule based expert system for rapid prototyping selection, *Robotics and Computer Integrated Manufacturing*, Vol. 18, pp. 267–274.

Appendix A: Glossary of Terms

ABS (acrylonitryl butadiene styrene): Thermoplastic RP material for Stratasys FDM machine.

AM (additive manufacturing): Additive manufacturing is the broader term for rapid prototyping or 3D printing that makes three-dimensional parts by adding materials one layer at a time.

BASS (break away support system): It is a system to remove the support material from FDM machine.

BT (build time): It is the time required to build a part using a 3D printer.

CAD (computer-aided design): CAD is the "front-end" for most rapid prototyping systems.

CAE (computer-aided engineering): CAE software allows for engineering analysis that includes determining a design's structural integrity or its heat transfer capacity.

CAM (computer-aided manufacturing): CAM generally refers to a system that uses CAD surface data to drive CNC machines, such as mills and lathes, in order to create parts, molds, and dies.

CNC (computer numerical control): Mills, lathes, and flame cutters are machines that are equipped with CNC capabilities.

CT (Computed tomography): It is a technique based on X-ray volumetric scanning of objects, used for medical and industrial purposes.

DOE (design of experiments): It is a statistical tool to improve a process.

EDM (electrical discharge machining): It is a machining process where an electrical spark is created between an electrode and a workpiece. This electric spark produces intense heat with temperatures reaching 8000°C–12000°C, melting almost anything.

FACET: Three- or four-sided polygon elements that represent a piece of a 3D polygonal mesh surface of the model. Triangular facets are used in STL files.

FDM (fused deposition modeling): In this method, a model created in a CAD program is imported into a software program specifically designed to work with the FDM machine

FFF (free-form fabrication): A more descriptive name for rapid prototyping methods.

GARPA (Global Alliance of Rapid Prototyping Association): It is an alliance between different rapid prototyping associations from around the world that fosters information transfer relating to rapid prototyping technology.

IGES (Initial Graphics Exchange Specifications): IGES is a standard industry format for exchanging CAD data between systems.

MEMS (microelectromechanical systems): It is the integration of mechanical elements: sensors, actuators, and electronics on a common silicon substrate through microfabrication technology.

MIMICS (materialise's interactive medical image control system): It is a software system that interfaces between scanner data and 3D printing data files for medical applications.

MJM (multi-jet modeling): 3D Systems Company uses inkjet technology to deposit materials for solid models.

MRI (magnetic resonance imaging): It is a technique that uses magnets to align electrons to create computer image that is used to create 3D file for medical applications.

LENS (laser engineered net shape): It is one of the first direct metal rapid prototyping system where the final parts are full-strength metals.

LOM (laminated object manufacturing): This process generates a part by laser-trimming materials to a desired cross-section. The material is delivered in sheet form. The cross-sections cut from the sheets are laminated into a solid block form using a thermal adhesive coating.

LS (laser sintering): It is a 3D printing process that uses energy generated by laser power to sinter powdered materials to make solid objects.

PJ (poly jet): It is a 3D printing process from originally Objet Geometries, now Stratasys, that deposits photocurable materials through an inkjet printer.

PLA (polylactic acid): It is a plant-based, biodegradable plastic material that is very popular with 3D printing machines.

PT (prototype tooling): Molds, dies, and other devices used to produce prototypes are sometimes referred to as soft tooling.

RE (reverse engineering): The science of taking an existing physical model and reproducing its surface geometry in a 3D data file on a CAD system.

RIM (reaction injection molding): This process uses a simple resin injection system with two pressurized chambers

RM (rapid manufacturing): It is similar to rapid prototyping which results in the production of end-use parts.

RP (rapid prototyping): A layer-by-layer additive process driven by the computer model data joins liquid, powder, or sheet materials to create free-form objects in plastic, wood, ceramic, metal, or composite materials.

RT (rapid tooling): Tooling that is driven from an RP process—the key to making it rapid. An approach to RT includes producing tooling components, such as mold inserts, directly from an RP process.

SB (service bureau): A service bureau is a company that owns and operates one or more rapid prototyping machines and will contract with other companies to use those machines to produce prototypes for them.

SLA (stereolithography apparatus): It is the first series of rapid prototyping machines made by 3D Systems.

SLS (selective laser sintering): First, a part is built by sintering when a laser beam hits a thin layer of powder material. Second, the part is built layer by layer. The next layer of powder is deposited by a roller mechanism on top of the previously formed layer. This powder layer is then sintered onto the previous layer by the laser.

SM (solid model): A 3D CAD model defined using solid modeling techniques. Solid models are preferred over surface models for RP because they define a closed, "water-tight" volume, a requirement of most RP systems.

STL (stereolithography): STL file formats are used to convert 3D CAD model data to physical parts using RP systems. The STL format, which is available in binary and ASCII form, uses triangular facets to approximate the shape of an object.

SM (surface model): A 3D CAD model defined by surfaces using a precise mathematical description such as Bezier B-spline surfaces or non-uniform rational B-spline surfaces or nonuniform rational B-spline (NURBS) surfaces. Surface models may or may not represent a closed volume.

Tooling: Molds, dies, and other devices for applications such as plastic injection molding, die casting, and sheet metal stamping.

UV (ultraviolet ray): It is an electromagnetic radiation that is used as energy source for making 3D printed parts and/or fabricating MEMS devices.

3DP (3D printing): A low-cost variation of an RP that is generally more efficient, user-friendly, and less expensive.

Appendix B: List of Abbreviations

3D	Three-dimensional
3DP	Three-dimensional printing
BPM	Ballistic particle manufacturing
CAD	Computer-aided design
CAE	Computer-aided engineering
CAM	Computer-aided manufacturing
CD	Computer disk
CIM	Computer integrated manufacturing
CLI	Common layer interface
CMM	Coordinate measuring machine
CNC	Computer numerical control
CT	Computed tomography
DOE	Design of experiments
DSPC	Direct shell production casting
EDM	Electric discharge machine
FDM	Fused deposition modeling
FEA	Finite element analysis
FEM	Finite element method
FFF	Free-form fabrication
GARPA	Global alliance of rapid prototyping association
HPGL	Hewlett-Packard graphics language
IGES	Initial graphics exchange specifications
LEAF	Layer exchange ASCII format
LENS	Laser engineered net shape
LMT	Layer manufacturing technologies
LOM	Laminated object manufacturing
LS	Laser sintering
MEMS	Micro-electro mechanical system
MIMICS	Materialise's interactive medical image control system
MJM	Multi-jet modeling
MJS	Multiphase jet solidification
MRI	Magnetic resonance imaging
NC	Numerical control
PJ	Poly jet
PLA	Polylactic acid
RE	Reverse engineering
RIM	Reaction injection molding
RM	Rapid manufacturing
RP	Rapid prototyping
RPI	Rapid prototyping interface

RPS	Rapid prototyping systems
RPT	Rapid prototyping technologies
RT	Rapid tooling
SAHP	Selective adhesive and hot press
SCS	Solid creation system
SFF	Solid freeform fabrication
SFM	Solid freeform manufacturing
SGC	Solid ground curing
SLA	Stereolithography apparatus
SLC	Stereolithography contour
SLS	Selective laser sintering
SOUP	Solid object ultraviolet-laser plotting
STL	Stereolithography file
UV	Ultraviolet ray

Appendix C: MATLAB Program for Expert System

```
%% Determine whether you are looking for an RP or 3DP

type=input('Are you looking for a Rapid Prototyping system or
a 3D Printer? For RP, please enter 1. For 3DP, please enter
2): ');

if type == 2

%% Expert system for 3D Printers

%% User inputs system specifications depending on the
parameters asked and user scales them out from 1 (lowest) to
10 (highest)

l=input('Please enter the minimum build platform length
desired (cm): ');
ll=input ('In scale of 1 (lowest) to 10 (highest), what
weight would you give the previous parameter? ');
w=input('Please enter the minimum build platform width
desired (cm): ');
ww=input ('In scale of 1 (lowest) to 10 (highest), what
weight would you give the previous parameter? ');
h=input('Please enter the minimum build platform height
desired (cm): ');
hh=input ('In scale of 1 (lowest) to 10 (highest), what
weight would you give the previous parameter? ');
p=input('Please enter the maximum price (USD): ');
pp=input ('In scale of 1 (lowest) to 10 (highest), what
weight would you give the previous parameter? ');
r=input('Please enter the maximum desired surface roughness
(resolution) (um): ');
rr=input ('In scale of 1 (lowest) to 10 (highest), what
weight would you give the previous parameter? ');
L=input('Please enter the maximum desired system physical
length (cm): ');
LL=input ('In scale of 1 (lowest) to 10 (highest), what
weight would you give the previous parameter? ');
W=input('Please enter the maximum desired system physical
width (cm): ');
```

```
WW=input ('In scale of 1 (lowest) to 10 (highest), what
weight would you give the previous parameter? ');
H=input('Please enter the maximum desired system physical
height (cm): ');
HH=input ('In scale of 1 (lowest) to 10 (highest), what
weight would you give the previous parameter? ');
z=input('Please enter the maximum desired system weight
(kg): ');
zz=input ('In scale of 1 (lowest) to 10 (highest), what
weight would you give the previous parameter? ');
s=input('Please enter the minimum desired speed of system
(mm/sec): ');
ss=input ('In scale of 1 (lowest) to 10 (highest), what
weight would you give the previous parameter? ');


%% system storage of 3D printers data

al=20.3;
aw=15;
ah=20.3;
ap=999;
ar=50;
aL=48.26;
aW=40.64;
aH=50.8;
az=7.25;
as=80;

bl=14;
bw=15;
bh=12.5;
bp=799;
br=50;
bL=30.48;
bW=30.48;
bH=30.48;
bz=2.72;
bs=100;

cl=22.3;
cw=22.3;
ch=20.5;
cp=2499;
cr=20;
cL=49.4;
cW=34.2;
cH=58.8;
cz=11.3;
cs=30;
```

```
dl=24.6;
dw=15.2;
dh=15.5;
dp=2499;
dr=100;
dL=49;
dW=42;
dH=53.1;
dz=12.6;
ds=135;

el=13.97;
ew=13.97;
eh=13.97;
ep=1299;
er=67;
eL=25.4;
eW=25.4;
eH=33;
ez=4.3;
es=20;

fl=12.5;
fw=12.5;
fh=16.5;
fp=1999;
fr=25;
fL=30;
fW=28;
fH=45;
fz=8;
fs=20;

gl=28;
gw=28;
gh=25;
gp=2500;
gr=80;
gL=66;
gW=52;
gH=52;
gz=11;
gs=200;

%% Eliminations of systems; all systems that satisfy the input
would receive one point times its scale


a=0;
b=0;
```

```
c=0;
d=0;
e=0;
f=0;
g=0;


if l <= al
    a=a+ll;
end
if l <= bl
    b=b+ll;
end
if l <= cl
    c=c+ll;
end
if l <= dl
    d=d+ll;
end
if l <= el
    e=e+ll;
end
if l <= fl
    f=f+ll;
end
if l <= gl
    g=g+ll;
end


if w <= aw
    a=a+ww;
end
if w <= bw
    b=b+ww;
end
if w <= cw
    c=c+ww;
end
if w <= dw
    d=d+ww;
end
if w <= ew
    e=e+ww;
end
if w <= fw
    f=f+ww;
end
if w <= gw
    g=g+ww;
```

```
end

if h <= ah
    a=a+hh;
end
if h <= bh
    b=b+hh;
end
if h <= ch
    c=c+hh;
end
if h <= dh
    d=d+hh;
end
if h <= eh
    e=e+hh;
end
if h <= fh
    f=f+hh;
end
if h <= gh
    g=g+hh;
end


if p >= ap
    a=a+pp;
end
if p >= bp
    b=b+pp;
end
if p >= cp
    c=c+pp;
end
if p >= dp
    d=d+pp;
end
if p >= ep
    e=e+pp;
end
if p >= fp
    f=f+pp;
end
if p >= gp
    g=g+pp;
end


if r >= ar
    a=a+rr;
```

```
end
if r >= br
    b=b+rr;
end
if r >= cr
    c=c+rr;
end
if r >= dr
    d=d+rr;
end
if r >= er
    e=e+rr;
end
if r >= fr
    f=f+rr;
end
if r >= gr
    g=g+rr;
end


if L >= aL
    a=a+LL;
end
if L >= bL
    b=b+LL;
end
if L >= cL
    c=c+LL;
end
if L >= dL
    d=d+LL;
end
if L >= eL
    e=e+LL;
end
if L >= fL
    f=f+LL;
end
if L >= gL
    g=g+LL;
end


if W >= aW
    a=a+WW;
end
if W >= bW
    b=b+WW;
end
```

```
if W >= cW
    c=c+WW;
end
if W >= dW
    d=d+WW;
end
if W >= eW
    e=e+WW;
end
if W >= fW
    f=f+WW;
end
if W >= gW
    g=g+WW;
end


if H >= aH
    a=a+HH;
end
if H >= bH
    b=b+HH;
end
if H >= cH
    c=c+HH;
end
if H >= dH
    d=d+HH;
end
if H >= eH
    e=e+HH;
end
if H >= fH
    f=f+HH;
end
if H >= gH
    g=g+HH;
end


if z >= az
    a=a+zz;
end
if z >= bz
    b=b+zz;
end
if z >= cz
    c=c+zz;
end
```

```
if z >= dz
    d=d+zz;
end
if z >= ez
    e=e+zz;
end
if z >= fz
    f=f+zz;
end
if z >= gz
    g=g+zz;
end

if s <= as
    a=a+ss;
end
if s <= bs
    b=b+ss;
end
if s <= cs
    c=c+ss;
end
if s <= ds
    d=d+ss;
end
if s <= es
    e=e+ss;
end
if s <= fs
    f=f+ss;
end
if s <= gs
    g=g+ss;
end


%% Calculate which system got the highest number of points

 X=[a b c d e f g];
 F=max(X);
 fprintf(1, '\n');
fprintf(1, '\n');

 if F == a
     disp('The Printerbot Simple 2016 best fits your needs.');
end

if F==b
    disp('The Bukito best fits your needs.');
end
```

```
if F==c
    disp('The Ultimaker 2+ best fits your needs.');
end

if F==d
    disp('The MakerBot Replicator 2x best fits your
needs.');
end

if F==e
    disp('The Cube 2 best fits your needs.');
end

if F == f
    disp('The Formlabs Form 1+ best fits your needs.');
end

if F == g
    disp('The Lulzbot Taz 6 best fits your needs.');
end



%% Display what parameter(s) was(were) not fulfilled

disp('The specification(s) that was/were not fulfilled is/
are:');

if max(X) == a
if l > al
    disp('platform build length');
end
if w > aw
    disp('platform build width');
end
if h > ah
    disp('platform build height');;
end
if p < ap
    disp('price');
end
if r < ar
    disp('surface roughness (resolution)');
end
if L < aL
    disp('system length');
end
if W < aW
    disp('system width');
end
```

```
if H < aH
    disp('system height');
end
if z < az
    disp('system weight');
end
if s > as
    disp('system speed');
end
end


if max(X) == b
if l > bl
    disp('platform build length');
end
if w > bw
    disp('platform build width');
end
if h > bh
    disp('platform build height');;
end
if p < bp
    disp('price');
end
if r < br
    disp('surface roughness (resolution)');
end
if L < bL
    disp('system length');
end
if W < bW
    disp('system width');
end
if H < bH
    disp('system height');
end
if z < bz
    disp('system weight');
end
if s > bs
    disp('system speed');
end
end

if max(X) == c
if l > cl
    disp('platform build length');
end
if w > cw
```

```
        disp('platform build width');
end
if h > ch
    disp('platform build height');;
end
if p < cp
    disp('price');
end
if r < cr
    disp('surface roughness (resolution)');
end
if L < cL
    disp('system length');
end
if W < cW
    disp('system width');
end
if H < cH
    disp('system height');
end
if z < cz
    disp('system weight');
end
if s > cs
    disp('system speed');
end
end

if max(X) == d

if l > dl
    disp('platform build length');
end
if w > dw
    disp('platform build width');
end
if h > dh
    disp('platform build height');;
end
if p < dp
    disp('price');
end
if r < dr
    disp('surface roughness (resolution)');
end
if L < dL
    disp('system length');
end
if W < dW
    disp('system width');
```

```
end
if H < dH
    disp('system height');
end
if z < dz
    disp('system weight');
end
if s > ds
    disp('system speed');
end
end

if max(X) == e

if l > el
    disp('platform build length');
end
if w > ew
    disp('platform build width');
end
if h > eh
    disp('platform build height');;
end
if p < ep
    disp('price');
end
if r < er
    disp('surface roughness (resolution)');
end
if L < eL
    disp('system length');
end
if W < eW
    disp('system width');
end
if H < eH
    disp('system height');
end
if z < ez
    disp('system weight');
end
if s > es
    disp('system speed');
end
end

if max(X) == f


if l > fl
```

```
      disp('platform build length');
end
if w > fw
      disp('platform build width');
end
if h > fh
      disp('platform build height');;
end
if p < fp
      disp('price');
end
if r < fr
      disp('surface roughness (resolution)');
end
if L < fL
      disp('system length');
end
if W < fW
      disp('system width');
end
if H < fH
      disp('system height');
end
if z < fz
      disp('system weight');
end
if s > fs
      disp('system speed');
end
end


if max(X) == g

if l > gl
      disp('platform build length');
end
if w > gw
      disp('platform build width');
end
if h > gh
      disp('platform build height');;
end
if p < gp
      disp('price');
end
if r < gr
      disp('surface roughness (resolution)');
end
if L < gL
```

```
    disp('system length');
end
if W < gW
    disp('system width');
end
if H < gH
    disp('system height');
end
if z < gz
    disp('system weight');
end
if s > gs
    disp('system speed');
end
end


fprintf(1, '\n');
second=input('Do you want to see the second best (For yes,
please enter 1. For no, please enter 2)? ');
fprintf(1, '\n');

if second == 1
F3=max(X(X~=max(X)));

if F3 == a
     disp('The Printerbot Simple 2016 comes in the second
place.');
end

if F3 ==b
    disp('The Bukito comes in the second place.');
end

if F3==c
    disp('The Ultimaker 2+ comes in the second place.');
end

if F3==d
    disp('The MakerBot Replicator 2x comes in the second
place.');
end

if F3==e
    disp('The Cube 2 comes in the second place.');
end

if F3 == f
    disp('The Formlabs Form 1+ comes in the second place.');
end
```

```
if F3 == g
    disp('The Lulzbot Taz 6 comes in the second place.');
end
elseif second == 2
        disp('OK! As you like!');
end

if second == 1
disp('The specification(s) that was/were not fulfilled by the
second best system is/are:');

if max(X(X~=max(X))) == a
if l > al
    disp('platform build length');
end
if w > aw
    disp('platform build width');
end
if h > ah
    disp('platform build height');;
end
if p < ap
    disp('price');
end
if r < ar
    disp('surface roughness (resolution)');
end
if L < aL
    disp('system length');
end
if W < aW
    disp('system width');
end
if H < aH
    disp('system height');
end
if z < az
    disp('system weight');
end
if s > as
    disp('system speed');
end
end

if max(X(X~=max(X))) == b
if l > bl
    disp('platform build length');
end
if w > bw
    disp('platform build width');
```

```
end
if h > bh
    disp('platform build height');;
end
if p < bp
    disp('price');
end
if r < br
    disp('surface roughness (resolution)');
end
if L < bL
    disp('system length');
end
if W < bW
    disp('system width');
end
if H < bH
    disp('system height');
end
if z < bz
    disp('system weight');
end
if s > bs
    disp('system speed');
end
end

if max(X(X~=max(X))) == c
if l > cl
    disp('platform build length');
end
if w > cw
    disp('platform build width');
end
if h > ch
    disp('platform build height');;
end
if p < cp
    disp('price');
end
if r < cr
    disp('surface roughness (resolution)');
end
if L < cL
    disp('system length');
end
if W < cW
    disp('system width');
end
if H < cH
```

```
        disp('system height');
    end
    if z < cz
        disp('system weight');
    end
    if s > cs
        disp('system speed');
    end
end

if max(X(X~=max(X))) == d

    if l > dl
        disp('platform build length');
    end
    if w > dw
        disp('platform build width');
    end
    if h > dh
        disp('platform build height');;
    end
    if p < dp
        disp('price');
    end
    if r < dr
        disp('surface roughness (resolution)');
    end
    if L < dL
        disp('system length');
    end
    if W < dW
        disp('system width');
    end
    if H < dH
        disp('system height');
    end
    if z < dz
        disp('system weight');
    end
    if s > ds
        disp('system speed');
    end
end

if max(X(X~=max(X))) == e

    if l > el
        disp('platform build length');
    end
    if w > ew
```

```
    disp('platform build width');
end
if h > eh
    disp('platform build height');;
end
if p < ep
    disp('price');
end
if r < er
    disp('surface roughness (resolution)');
end
if L < eL
    disp('system length');
end
if W < eW
    disp('system width');
end
if H < eH
    disp('system height');
end
if z < ez
    disp('system weight');
end
if s > es
    disp('system speed');
end
end

if max(X(X~=max(X))) == f


if l > fl
    disp('platform build length');
end
if w > fw
    disp('platform build width');
end
if h > fh
    disp('platform build height');;
end
if p < fp
    disp('price');
end
if r < fr
    disp('surface roughness (resolution)');
end
if L < fL
    disp('system length');
end
if W < fW
```

```
        disp('system width');
end
if H < fH
        disp('system height');
end
if z < fz
        disp('system weight');
end
if s > fs
        disp('system speed');
end
end

if max(X(X~=max(X))) == g

if l > gl
        disp('platform build length');
end
if w > gw
        disp('platform build width');
end
if h > gh
        disp('platform build height');;
end
if p < gp
        disp('price');
end
if r < gr
        disp('surface roughness (resolution)');
end
if L < gL
        disp('system length');
end
if W < gW
        disp('system width');
end
if H < gH
        disp('system height');
end
if z < gz
        disp('system weight');
end
if s > gs
        disp('system speed');
end
end
end

end
```

```
%% Please enter data


if type == 1


%% Expert system for Rapid Prototyping

%% User inputs system specifications according to the
parameters shown and user scales them out from 1 (lowest) to
10 (highest)

l=input('Please enter the minimum build platform length
desired (cm): ');
ll=input ('In scale of 1 (lowest) to 10 (highest), what
weight would you give the previous parameter? ');
w=input('Please enter the minimum build platform width
desired (cm): ');
ww=input ('In scale of 1 (lowest) to 10 (highest), what
weight would you give the previous parameter? ');
h=input('Please enter the minimum build platform height
desired (cm): ');
hh=input ('In scale of 1 (lowest) to 10 (highest), what
weight would you give the previous parameter? ');
p=input('Please enter the maximum price (USD): ');
pp=input ('In scale of 1 (lowest) to 10 (highest), what
weight would you give the previous parameter? ');
r=input('Please enter the maximum desired surface roughness
(resolution)(um): ');
rr=input ('In scale of 1 (lowest) to 10 (highest), what
weight would you give the previous parameter? ');
L=input('Please enter the maximum desired system length
(cm): ');
LL=input ('In scale of 1 (lowest) to 10 (highest), what
weight would you give the previous parameter? ');
W=input('Please enter the maximum desired system width
(cm): ');
WW=input ('In scale of 1 (lowest) to 10 (highest), what
weight would you give the previous parameter? ');
H=input('Please enter the maximum desired system height
(cm): ');
HH=input ('In scale of 1 (lowest) to 10 (highest), what
weight would you give the previous parameter? ');
z=input('Please enter the maximum desired system weight
(kg): ');
zz=input ('In scale of 1 (lowest) to 10 (highest), what
weight would you give the previous parameter? ');
```

```
%% system storage of RP data

al=10;
aw=10;
ah=9.5;
ap=100000;
ar=40;
aL=80;
aW=95;
aH=225;
az=580;

bl=40.6;
bw=35.5;
bh=40.6;
bp=100000;
br=200;
bL=129.5;
bW=90.2;
bH=198.4;
bz=680;

cl=49;
cw=39;
ch=20;
cp=100000;
cr=20;
cL=140;
cW=126;
cH=110;
cz=430;

dl=26.6;
dw=17.5;
dh=19.3;
dp=50000;
dr=100;
dL=74;
dW=76;
dH=117;
dz=90;

el=25;
ew=25;
eh=25;
ep=200000;
er=100;
eL=78.7;
```

```
eW=73.7;
eH=182.9;
ez=181;

fl=35.5;
fw=30.5;
fh=30.5;
fp=100000;
fr=200;
fL=129.5;
fW=90.2;
fH=198.4;
fz=680;

gl=10;
gw=10;
gh=10;
gp=299000;
gr=25;
gL=100;
gW=100;
gH=150;
gz=200;



%% Eliminations of systems; all systems that satisfy the input
would receive one point times its scale


a=0;
b=0;
c=0;
d=0;
e=0;
f=0;
g=0;

if 1 <= al
    a=a+ll;
end
if 1 <= bl
    b=b+ll;
end
if 1 <= cl
    c=c+ll;
end
if 1 <= dl
    d=d+ll;
end
```

```
if l <= el
    e=e+ll;
end
if l <= fl
    f=f+ll;
end
if l <= gl
    g=g+ll;
end

if w <= aw
    a=a+ww;
end
if w <= bw
    b=b+ww;
end
if w <= cw
    c=c+ww;
end
if w <= dw
    d=d+ww;
end
if w <= ew
    e=e+ww;
end
if w <= fw
    f=f+ww;
end
if w <= gw
    g=g+ww;
end

if h <= ah
    a=a+hh;
end
if h <= bh
    b=b+hh;
end
if h <= ch
    c=c+hh;
end
if h <= dh
    d=d+hh;
end
if h <= eh
    e=e+hh;
end
if h <= fh
    f=f+hh;
end
```

```
if h <= gh
    g=g+hh;
end


if p >= ap
    a=a+pp;
end
if p >= bp
    b=b+pp;
end
if p >= cp
    c=c+pp;
end
if p >= dp
    d=d+pp;
end
if p >= ep
    e=e+pp;
end
if p >= fp
    f=f+pp;
end
if p >= gp
    g=g+pp;
end


if r >= ar
    a=a+rr;
end
if r >= br
    b=b+rr;
end
if r >= cr
    c=c+rr;
end
if r >= dr
    d=d+rr;
end
if r >= er
    e=e+rr;
end
if r >= fr
    f=f+rr;
end
if r >= gr
    g=g+rr;
end
```

```
if L >= aL
    a=a+LL;
end
if L >= bL
    b=b+LL;
end
if L >= cL
    c=c+LL;
end
if L >= dL
    d=d+LL;
end
if L >= eL
    e=e+LL;
end
if L >= fL
    f=f+LL;
end
if L >= gL
    g=g+LL;
end


if W >= aW
    a=a+WW;
end
if W >= bW
    b=b+WW;
end
if W >= cW
    c=c+WW;
end
if W >= dW
    d=d+WW;
end
if W >= eW
    e=e+WW;
end
if W >= fW
    f=f+WW;
end
if W >= gW
    g=g+WW;
end


if H >= aH
    a=a+HH;
end
if H >= bH
```

```
     b=b+HH;
end
if H >= cH
     c=c+HH;
end
if H >= dH
     d=d+HH;
end
if H >= eH
     e=e+HH;
end
if H >= fH
     f=f+HH;
end
if H >= gH
     g=g+HH;
end


if z >= az
     a=a+zz;
end
if z >= bz
     b=b+zz;
end
if z >= cz
     c=c+zz;
end
if z >= dz
     d=d+zz;
end
if z >= ez
     e=e+zz;
end
if z >= fz
     f=f+zz;
end
if z >= gz
     g=g+zz;
end


%% Calculate which system got the highest number of points

 X=[a b c d e f g];
 F=max(X);
 fprintf(1, '\n');
fprintf(1, '\n');

 if F == a
```

```matlab
        disp('The EOS M 100 best fits your needs.');
end

if F==b
    disp('The Stratasys Fortus 450mc best fits your needs.');
end

if F==c
    disp('The Stratasys Object 500 Connex3 best fits your
needs.');
end

if F==d
    disp('The EnvisionTec ULTRA 3SP best fits your needs.');
end

if F==e
    disp('The 3DS ProJet 6000 SD best fits your needs.');
end

if F == f
    disp('The Stratasys Fortus 380mc best fits your needs.');
end

if F == g
    disp('The Optomec LENS 450 best fits your needs.');
end


%% Display what parameter was not fulfilled

disp('The specification(s) that was/were not fulfilled is/
are:');

if max(X) == a
if l > al
    disp('platform build length');
end
if w > aw
    disp('platform build width');
end
if h > ah
    disp('platform build height');;
end
if p < ap
    disp('price');
end
if r < ar
    disp('surface roughness');
end
```

```
if L < aL
    disp('system length');
end
if W < aW
    disp('system width');
end
if H < aH
    disp('system height');
end
if z < az
    disp('system weight');
end
end

if max(X) == b
if l > bl
    disp('platform build length');
end
if w > bw
    disp('platform build width');
end
if h > bh
    disp('platform build height');;
end
if p < bp
    disp('price');
end
if r < br
    disp('surface roughness');
end
if L < bL
    disp('system length');
end
if W < bW
    disp('system width');
end
if H < bH
    disp('system height');
end
if z < bz
    disp('system weight');
end
end

if max(X) == c
if l > cl
    disp('platform build length');
end
if w > cw
    disp('platform build width');
```

```
end
if h > ch
    disp('platform build height');;
end
if p < cp
    disp('price');
end
if r < cr
    disp('surface roughness');
end
if L < cL
    disp('system length');
end
if W < cW
    disp('system width');
end
if H < cH
    disp('system height');
end
if z < cz
    disp('system weight');
end
end

if max(X) == d

if l > dl
    disp('platform build length');
end
if w > dw
    disp('platform build width');
end
if h > dh
    disp('platform build height');;
end
if p < dp
    disp('price');
end
if r < dr
    disp('surface roughness');
end
if L < dL
    disp('system length');
end
if W < dW
    disp('system width');
end
if H < dH
    disp('system height');
end
```

```matlab
if z < dz
    disp('system weight');
end
end

if max(X) == e

if l > el
    disp('platform build length');
end
if w > ew
    disp('platform build width');
end
if h > eh
    disp('platform build height');;
end
if p < ep
    disp('price');
end
if r < er
    disp('surface roughness');
end
if L < eL
    disp('system length');
end
if W < eW
    disp('system width');
end
if H < eH
    disp('system height');
end
if z < ez
    disp('system weight');
end
end

if max(X) == f
if l > fl
    disp('platform build length');
end
if w > fw
    disp('platform build width');
end
if h > fh
    disp('platform build height');;
end
if p < fp
    disp('price');
end
if r < fr
```

```
        disp('surface roughness');
end
if L < fL
    disp('system length');
end
if W < fW
    disp('system width');
end
if H < fH
    disp('system height');
end
if z < fz
    disp('system weight');
end
end

if max(X) == g

if l > gl
    disp('platform build length');
end
if w > gw
    disp('platform build width');
end
if h > gh
    disp('platform build height');;
end
if p < gp
    disp('price');
end
if r < gr
    disp('surface roughness');
end
if L < gL
    disp('system length');
end
if W < gW
    disp('system width');
end
if H < gH
    disp('system height');
end
if z < gz
    disp('system weight');
end
end

fprintf(1, '\n');
second=input('Do you want to see the second best (For yes,
please enter 1. For no, please enter 2)? ');
```

```
fprintf(1, '\n');

if second == 1
F3=max(X(X~=max(X)));

if F3 == a
    disp('The EOS M 100 comes in the second place.');
end

if F3 ==b
    disp('The Stratasys Fortus 450mc comes in the second
place.');
end

if F3==c
    disp('The Stratasys Object 500 Connex3 comes in the second
place.');
end

if F3==d
    disp('The EnvisionTec ULTRA 3SP comes in the second
place.');
end

if F3==e
    disp('The 3DS ProJet 6000 SD comes in the second place.');
end

if F3 == f
    disp('The Stratasys Fortus 380mc comes in the second
place.');
end

if F3 == g
    disp('The Optomec LENS 450 comes in the second place.');
end
elseif second == 2
        disp('OK! As you like!');
end

if second == 1

disp('The specification(s) that was/were not fulfilled by the
second best system is/are:');

if max(X(X~=max(X))) == a
if l > al
    disp('platform build length');
end
if w > aw
```

```
        disp('platform build width');
end
if h > ah
        disp('platform build height');;
end
if p < ap
        disp('price');
end
if r < ar
        disp('surface roughness');
end
if L < aL
        disp('system length');
end
if W < aW
        disp('system width');
end
if H < aH
        disp('system height');
end
if z < az
        disp('system weight');
end
end

if max(X(X~=max(X))) == b
if l > bl
        disp('platform build length');
end
if w > bw
        disp('platform build width');
end
if h > bh
        disp('platform build height');;
end
if p < bp
        disp('price');
end
if r < br
        disp('surface roughness');
end
if L < bL
        disp('system length');
end
if W < bW
        disp('system width');
end
if H < bH
        disp('system height');
end
```

```matlab
if z < bz
    disp('system weight');
end
end

if max(X(X~=max(X))) == c
if l > cl
    disp('platform build length');
end
if w > cw
    disp('platform build width');
end
if h > ch
    disp('platform build height');;
end
if p < cp
    disp('price');
end
if r < cr
    disp('surface roughness');
end
if L < cL
    disp('system length');
end
if W < cW
    disp('system width');
end
if H < cH
    disp('system height');
end
if z < cz
    disp('system weight');
end
end

if max(X(X~=max(X))) == d

if l > dl
    disp('platform build length');
end
if w > dw
    disp('platform build width');
end
if h > dh
    disp('platform build height');;
end
if p < dp
    disp('price');
end
```

```
if r < dr
    disp('surface roughness');
end
if L < dL
    disp('system length');
end
if W < dW
    disp('system width');
end
if H < dH
    disp('system height');
end
if z < dz
    disp('system weight');
end
end

if max(X(X~=max(X))) == e

if l > el
    disp('platform build length');
end
if w > ew
    disp('platform build width');
end
if h > eh
    disp('platform build height');;
end
if p < ep
    disp('price');
end
if r < er
    disp('surface roughness');
end
if L < eL
    disp('system length');
end
if W < eW
    disp('system width');
end
if H < eH
    disp('system height');
end
if z < ez
    disp('system weight');
end
end

if max(X(X~=max(X))) == f
```

```
 if l > fl
    disp('platform build length');
end
if w > fw
    disp('platform build width');
end
if h > fh
    disp('platform build height');;
end
if p < fp
    disp('price');
end
if r < fr
    disp('surface roughness');
end
if L < fL
    disp('system length');
end
if W < fW
    disp('system width');
end
if H < fH
    disp('system height');
end
if z < fz
    disp('system weight');
end
end

if max(X(X~=max(X))) == g

if l > gl
    disp('platform build length');
end
if w > gw
    disp('platform build width');
end
if h > gh
    disp('platform build height');;
end
if p < gp
    disp('price');
end
if r < gr
    disp('surface roughness');
end
if L < gL
    disp('system length');
end
if W < gW
```

```
    disp('system width');
end
if H < gH
    disp('system height');
end
if z < gz
    disp('system weight');
end
end
end
end


%% Please enter data
```

Index

Printed and bound by CPI Group (UK) Ltd, Croydon, CR0 4YY

01/11/2024

01782624-0002